PROPERTY OF FRED W. PRIOR

Form and Pattern in Human Evolution

Charles Oxnard

Form and Pattern in Human Evolution

Some Mathematical, Physical, and Engineering Approaches

The University of Chicago Press
Chicago & London

CHARLES E. OXNARD is professor in the Department of Anatomy, the Department of Anthropology, and the Committee on Evolutionary Biology at the University of Chicago. He is also master of the Biological Sciences Collegiate Division and dean of the College, the Division of Biological Sciences and the Pritzker School of Medicine. His list of scholarly publications totals over seventy.
[1973]

The University of Chicago Press, Chicago 60637

The University of Chicago Press, Ltd., London

© 1973 by The University of Chicago

All rights reserved. Published 1973

Printed in the United States of America

International Standard Book Number: 0-226-64251-8

Library of Congress Catalog Card Number: 72-89884

Contents

Preface vii

1. The Characterization and Comparison of Form and Pattern

The Shapes of Animals 1

Classical Methods for Studying Shape 3

Complex Metrical Analyses of Shape 6

Some Deficiencies of Mensurational Data 9

A Return to the Pictorial Data Set: Optical Methods 14

The Functional Role of Form 15

The Time Dimension: Fossils 16

2. Associations between Function and Morphology

Introduction 19

A Strategy of Investigation 19

Morphological Studies of the Primate Shoulder 26

Another Example: The Primate Hand 31

The Pelvis in Primates 33

Summary 36

3. Multivariate Morphometric Analyses

Introduction 37

The Structure of a Single Group 39

The Interrelationships of Many Groups 43

Biological Meaning of Mathematical Parameters 48

The Investigation of the Primate Shoulder: "Locomotor" Dimensions 50

Investigation of "Residual" Dimensions 51

Canonical Analysis of "Combined" Shoulder Dimensions 52

Study of the Shoulder in Nonprimate Mammals 56

A Return to the Original Dimensions 62

Conclusions on the Primate Shoulder 65

A Preliminary Study of the Primate Hip 67

4. Some Simple Testing Procedures

Introduction 69

The Technique of Data Collection 70

Biological Facets of the Data 71

Problems Inherent in the Nature of the Data Themselves 77

Conclusion 87

5. Group Finding Procedures in Morphology

Some Limitations of Multivariate Methods 88

Some Properties of Data Spaces 91

Group Finding Procedures 94

Application to the Shoulder in Primates: Evidence of Functional Information 99

Perception of Irregular Groups 105

Morphological Peaks and Troughs 109

The Combined Use of Multivariate and Cluster Analyses 111

A Note of Caution 118

6. Functional Significance of Bone Form: Experimental Stress Analysis

Introduction 122

A Description of Photoelastic Analysis 124

Stress Analysis of Bone Shapes 127

The Form of Hominoid Hand Bones 127

Theoretical Applications: Adaptation of Bones to Stress 135

7. Extrapolation to "Unknown" Data

Introduction 145

A Test with a "Living" Unknown 145

Some Fossil Unknowns: The Sterkfontein Innominate Bone 151

Fossil Fragments of the Shoulder Girdle 155

Other Interpolative Studies: Foot Bones 159

Neighborhood Limited Classification 163

Experimental Stress Analysis: The Olduvai Hand 164

A Speculative Conclusion 165

8. Optical Data Analysis in Morphology

Introduction 169

The Analysis of Patterns 171

Optical Data Analysis: The Basic Technique 176

An Example from Geology 181

Some Preliminary Investigations of Trabecular Pattern 183

Theoretical Applications to Problems of Bone Morphology 189

Further Speculation on Analysis of Form and Pattern in Biology 193

References 202
Author Index 213
Species Index 215
Subject Index 217

Preface

This book represents an attempt to provide, for those interested in the morphological evolution of monkeys, apes, and man, some insight into a number of mathematical, physical, and engineering approaches that may help in the study of form and pattern. These various techniques are not new; multivariate statistics and group-finding procedures have been available for several decades, and photoelasticity and Fourier optics derive originally from the nineteenth century. But the ability and desire of biologists to use these methods for the solution of genuine biological problems certainly is new. This ability has been conferred partly by the pioneering work of earlier investigators (in multivariate statistics, for example, by workers such as Fisher and Hotelling), partly by the evolution of instruments such as the electronic computer and the laser (which allow previously arduous investigations to be carried out with relative ease), and partly by an increased interest in morphological problems that has been engendered in so many fields outside biology.

For those who wish to utilize these techniques, adequate formal descriptions are available in the specialist literature. In addition such workers would do well to consult and collaborate closely with the appropriate mathematical and physical specialists. Certainly it is not the aim of this book to usurp these important functions.

What I wish to provide here are general nontechnical descriptions of the different methods so that those interested in modern ways of investigating the shape of primates are not forced to avoid a certain (and growing) portion of the literature. It is my hope to interest a wide range of morphologists (anatomists, anthropologists, zoologists), who may have no need or desire themselves to use these techniques, in understanding, assessing critically, and appraising those studies which do utilize such techniques. Thus the methods and results may be incorporated into scholarship and teaching.

It is especially my desire to present these materials to those students who already have interests in the study of shape and structure, but who do not wish to be limited by dissection and observation. Thus they may see that the investigation of the form of primates can go further and can reveal exciting possibilities not available to the unaided eye. These possibilities are evident not only from understanding the techniques themselves, but also from seeing the nature of the problems they are capable of tackling and the caliber of the results that flow from their use.

The various methods presented here are in different stages of development. The multivariate statistical work, for instance, has been underway since 1959. The studies utilizing optical data analysis, on the other hand, are very new and have scarcely passed the testing grade. Various other investigations are at intermediate levels. Accordingly, the presentations in the different chapters pass from those that are soundly based and fully documented to those that are unashamedly pilot and preliminary in nature. This is a deliberate attempt to place possible future directions of study before the reader.

If I manage to pass on to a few primatologists, anatomists, and biological anthropologists interested in the study of shape some enlightenment about the techniques that I have found valuable, some concept of the ways in which they may be used (research strategies and tactics), some information about actual results obtained that are not obviously available, and finally some idea of my own interest—indeed excitement—in these researches, then I shall consider the effort that has gone into this book to be very well directed.

The initial stimulus for this book and for its general form came from the 1968 Symposium at Burg Wartenstein on the Evolutionary Biology of Primates sponsored by the Wenner-Gren Foundation for Anthropological Research. The basic format of the book also owes much to its presentation to, and its reception by, those students at the University of Chicago who have taken my graduate courses in the analysis of form and pattern and in bone-joint-muscle biomechanics. Perhaps most of all, however, this book, along with the more recent developments of my studies, has been stimulated by my colleagues in evolutionary morphology in the Department of Anatomy at the University of Chicago. Professors Ronald Singer and Jack T. Stern, Jr., have participated in discussions and joint courses, have exchanged views about research strategies, and have criticized many manuscripts. In particular they have read several drafts of this work and are responsible for much improvement.

The broader background of research on which this book is based also owes much to collaborators and colleagues at the University of Birmingham in England and at the University of Kansas. Professor E. H. Ashton, Dr. R. M. Flinn, and Mr. T. F. Spence of the University of Birmingham have been and continue to be collaborators in much of my research. Professors P. M. Neely and J. C. Davis of the University of Kansas have also helped greatly with aspects of these studies.

It is particularly appropriate to acknowledge the stimulus and continuing interest and collaboration over the years provided by Professor Lord Zuckerman, OM, KCB, MD, DSc, FRS, previously Sands Cox Professor of Anatomy at the University of Birmingham.

It is also an especial pleasure to acknowledge the help and expertise of my personal research assistant, Miss Joan Hives, who has taken part in all my studies for several years and who has contributed greatly in the preparation of this book. She has willingly undertaken technical, bibliographic, artistic and drafting, computational, and secretarial tasks of all kinds. In addition she has attacked research problems and provided research ideas at many stages of these studies, especially in the area of experimental stress analysis.

Many other hands have contributed to this work through technical assistance in the laboratory, artistic and technical drawing, and heavy computational work. To be thanked in this regard are Mrs. Virginia Bates, Mrs. Marcia Greaves, Mrs. Cleone Hawkinson, Miss Susan Kirchmeyer, and Mrs. Rebecca Mead. The photographic expertise of Miss Shirley Aumiller, Mr. Melvin Oster, and Mr. Leslie A. Siemens is gratefully acknowledged. The biomedical staff of the

Joseph R. Regenstein Library at the University of Chicago has also been particularly helpful.

My work has been supported by United States Public Health Service grants HD 0054 and HD 02852, by National Science Foundation grant GS 30508, by grants from the Louis Block Bequest Fund and the Dr. W. C. and C. A. Abbott Memorial Fund of the University of Chicago, and by funds from the Wenner-Gren Foundation for Anthropological Research.

The following have kindly given permission for me to utilize photographs from original publications: plate 1, Academic Press, and Professor A. Rosenfeld and Mr. J. Strong; plate 8, H. C. Becker et al., *Annals of the New York Academy of Sciences* 57 (1969): 471–73, figs. 8, 9, 10, and 11, © 1969 by The New York Academy of Sciences, reprinted by permission; plate 9, Dr. Arnold R. Shulman, NASA, Goddard Space Flight Center; plate 10, Mr. Joel Gunn, University of Kansas, Lawrence; plates 11 and 12, Academic Press, and Professor H. C. Andrews; figure 126, the Jet Propulsion Laboratory, Pasadena, California, and Dr. R. H. Selzer; figure 127, © 1970 by John Wiley and Sons, and Dr. Arnold R. Shulman, NASA, Goddard Space Flight Center.

1   The Characterization and Comparison of Form and Pattern

The Shapes of Animals

The study of the shapes of animals has posed questions that have taxed men's minds since the earliest times. From an abstract interest in perfect shapes that is reflected in much preevolutionary work, the emphasis changed to supporting and then documenting evolution. In the present century the study of the shape of animals has become more and more oriented toward understanding, on the one hand, interactions of shape and function and, on the other, processes responsible for change in shape. Within both ontogeny and evolution these studies have progressed apace: we are now learning a great deal about molecular and epigenetic controls associated with differentiation, development, and growth, and about mechanism and process in evolution.

One area, however, that has not been well investigated is the analysis of change in shape itself. A few really great minds have been brought to bear upon the problems. D'Arcy Thompson's theory of transformations (1917, 1942), Fisher's discriminant functions (1936), Huxley and allometric growth (1932), Woodger and mapping techniques (1945), among others, readily spring to mind. But since those studies, the manner of investigating evolutionary change in shape has essentially remained static, though we recognize the fascinating but difficult theoretical work of Thom in ontogeny (1970) and the creative phylogenetic research technique of Sneath (1967). Especially has a lack of improvement been the case in studies of the morphological evolution of man and his closest living relatives, the monkeys and apes.

Although considerable advances in the understanding of human evolution have been made in recent years, they have rested, in the main, upon studies other than those of the evolution of shape. Biochemical, molecular, and genetic researches, naturally utilizing new methods, together with study, by traditional methods, of new materials (fossils), have provided the main insights.

Morphological investigations of extant primates and of fossils already well known have, in general, limited themselves to the self-sufficient techniques of dissection of soft tissues and observation of hard structure. Such findings as have been made thereby have certainly succeeded in providing new information; but this has usually resulted in the closure of relatively minor gaps in an already fairly well known pattern rather than in the creation of new vistas.

In some ways it is easy to see why this has been so. The earlier masters (Huxley, D'Arcy Thompson, etc.) include within their writings clear indication of the difficulties: essentially they had gone as far as was possible with the tools available to them. Just as microscopists awaited the electron microscope before being able to venture to the ultrastructural level, so morphologists needed new tools before venturing far from assessment of shape by observation (occasionally backed by measurements and simple analysis) towards more complex investigations of underlying factors of shape. However, the evolution of a number of modern tools (especially the electronic computer and computer software) provides mechanisms for new orders of investigation. It only remains for primate morphologists to use them.

Morphologists of other kinds (nonbiological morphologists) have not been slow to take up a variety of new methods. Geologists, geographers, astronomers, metallurgists, sociologists, psychologists, and others, all deal with problems of structure; and in many ways their problems are far more difficult than are those of the anatomist and anthropologist. A metallurgist's morphology, of course, is likely to be as easy to characterize as that of an anatomist; in fact in some ways it is easier, because he deals with inert materials. But surely the structural problems of the sociologist or economist are considerably more difficult to pin down than are those of the primate morphologist. The psychologist may have among the most difficult of all structural problems. Yet all these workers in their different disciplines have taken into their methodologies a whole range of techniques; essentially they have adopted anything that it seemed might help. In so doing they have sometimes undoubtedly run into cul-de-sacs; some methods may have turned out to be red herrings; others, however, have proven especially helpful.

This inventiveness has occasionally intruded into biology. This especially has been so in recent years in the discipline of taxonomy (Sokal and Sneath, 1963; Cole, 1969). At an even earlier date, those interested in specifically human variation and growth had already adopted a variety of techniques; this latter group has, among others, provided some of the most pioneering studies that have been carried out (e.g., Mukherjee, Rao, and Trevor, 1955). But in evolutionary investigations of the shape of primates, these advances have made scarcely any dent.

There is no doubt that the self-sufficiency of the techniques of dissection and observation has prevented researchers from utilizing new methods. Both of these older techniques are time-consuming and laborious, yet neither can be relegated to technical help; both require the complete attention, the experience, and the expertise of the researcher himself and leave little room for other technical experimentation.

It is also true that the early use of newer analytical methods was laborious. Much of this work has been rejected as mere number grinding; some anatomists and anthropologists still look upon quantitative morphological studies as extravagant expenditures of time and man power, often for results that seem to add little to our knowledge. There can be no doubt that criticism of this type is superficial; without the laborious but pioneering studies, often rendered more difficult by the lack of techniques and equipment nowadays regarded as indispensable, it would not have been possible to go beyond the confines provided by interpretation based upon personal observation and dissection. For it has only been the use of currently available formulae to the extreme of their capabilities that has conferred upon researchers the competence to help propose features and criteria that may be used for the development of yet better approaches.

But I think that one of the chief stumbling blocks in the adoption of newer techniques lies in what has been thought to be the difficult nature of the methods, and in a lack of understanding of the extent to which they may be valuable.

Thus it is that, having myself attempted to apply a number of these ways of looking at evolutionary differences in the shape of primates, I have been led to some understanding of how they work and of what new insights they are capable of providing. A sufficient number of studies have now been carried out that some of the research strategies and tactics involved can be distinguished.

Of course, all of these research methods are best described technically by workers in those fields from which the primary development springs. It is to the statistician, the physicist, the computer expert, that we must turn if we wish to use these methods. Such elucidation is already available in the literature; much research is going on for further discovery and development.

My role here is not that of the manufacturer of techniques, but of the consumer; my brief in this text is to attempt to describe (a) the nature of these various strategies, (b) the manner in which they may be used by morphologists, and (c) some indication of the advances in understanding of the evolution of morphology in primates through the results that they provide. Within that brief my aim is to attempt so to make these descriptions that special background is not necessary for their understanding.

Classical Methods for Studying Shape

The chief agent for studying shape in primates is the experienced and creative human mind behind the human eye. This equipment, with its exquisite powers of recognition and discrimination, has provided the great bulk of what we now know about primate morphology. Its power of discrimination and recognition is evident in the puzzles provided in figures 1 and 2. And in the comparison of the form of bones, similar powers are evident.

Thus we may note the size and shape of bones and bony fragments; long bones may be compared with short ones, and differences in proportion may be distinguished between one bone and the next. Further, we may observe architecture: the positions and orientations of bony buttresses and the changing patterns of trabecular and compact bone are readily apparent to the unassisted eye; the relationships of osteones with one another, and intraosteone structure, are revealed by the microscope; even at the ultrastructural grade, pictorial information is available about the interrelationships of, for instance, collagen and hydroxyapatite. Such architectural data often exist as two-dimensional slices ranging from a photograph of a macroscopic wafer of bone to electron micrographs of ultrafine sections or of surfaces. Detailed inspection of serial sections, together with three-dimensional methods (e.g., the use of x-rays, especially stereoscopic x-rays) are allowing some understanding of three-dimensional aspects of the morphology. At all levels, however, comprehension is in general confined to major and usually obvious patterns (for instance, those of principal osteone directions, or of major trabecular delineations). This relates inevitably to the pictorial nature of the evidence and the comparisons. The attempt to synthesize by eye all the information contained in the plane or surface view of an object, for comparison with many similar objects, results in the loss of much that is subtle.

Thus the observational method has its limitations, and these are well known. One of these relates to quantitative assessment. Although the mind can very

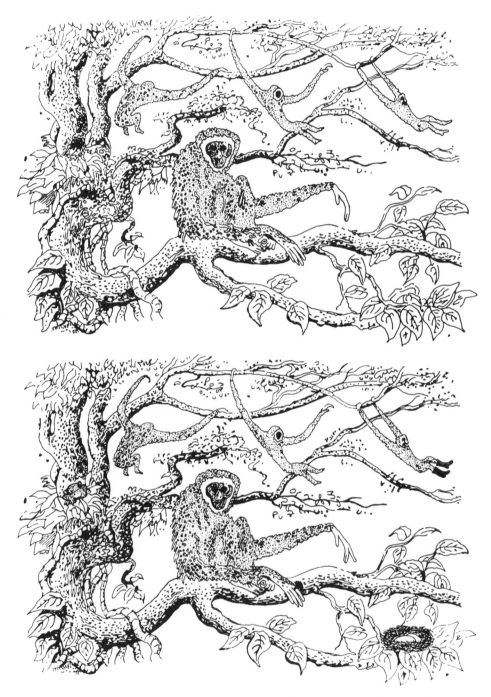

Figure 1. Powers of discrimination. In these two pictures (which are not identical, owing to the difficulty of identical reproduction by the artist) there have been introduced five deliberate mistakes. The human eye has no difficulty distinguishing deliberate differences from artistic license. It would be fairly difficult to program a computer to make these diagnoses.

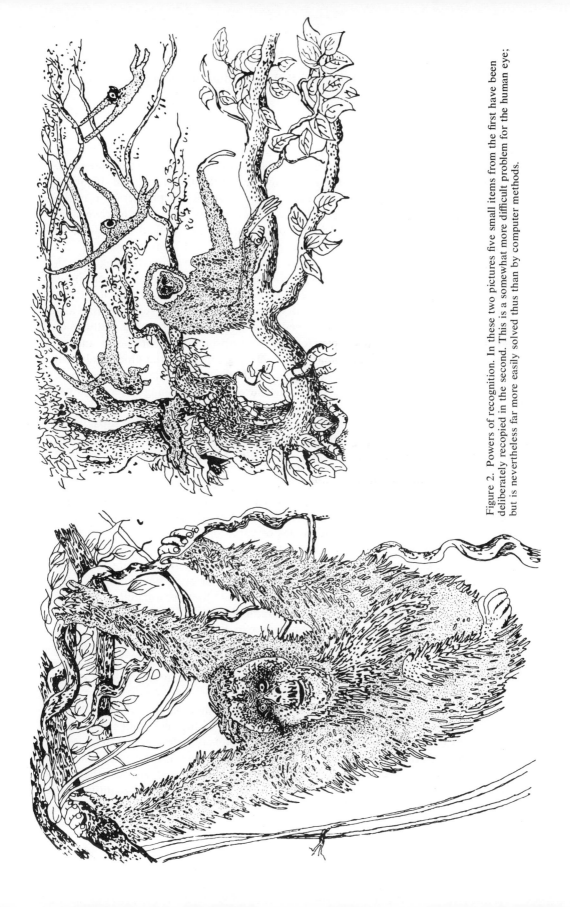

Figure 2. Powers of recognition. In these two pictures five small items from the first have been deliberately recopied in the second. This is a somewhat more difficult problem for the human eye; but is nevertheless far more easily solved thus than by computer methods.

readily discriminate objects of slightly differing sizes when these are lying side by side, it can do so less well if the objects have to be compared in the memory. The old techniques of measurement together with univariate and bivariate analyses, have been the traditional attempt to overcome this problem. From the earliest times mensurational methods have been utilized in order to reveal information contained within pictorial representations but difficult otherwise to elicit. For instance, absolute measurement is a clear aid to assessment of size differences, while the comparison of two sizes leads to an understanding of such properties as allometry and correlation, concepts difficult to determine pictorially. Combinations of measurements have resulted in the widespread use of indices (and angles, as disguised indices), and when it is difficult to separate size and shape effects in analysis, such combined dimensions are reasonable methods of allowing for differences in gross size. For both absolute measurements and indices, univariate and bivariate analyses have long remained the chief analytical tactics. Even today extensive studies (e.g., Tobias, 1967) utilize such methods.

An example of the power of simple measurement in defining differences between structures is the data by Dr. Anderson (described by Fisher, 1936) taken on a common flower, the iris. Here four measurements (of petal and sepal width and length) suffice to distinguish completely *Iris setosa* and to separate, albeit with some overlap, *I. virginica* from *I. versicolor*, as is shown in figure 3.

Complex Metrical Analyses of Shape

There remains a further deficiency that is shared by both observation and simple mensuration: namely, in the characterization and comparison of form, complex interrelationships such as differing modes of variation and varying kinds of multiple correlation may be present. In fact, biological shapes provide a series of highly complex examples of phenomena such as these. Multivariate statistical methods of various kinds are capable of allowing for such perturbations (e.g., variation and covariation) within extensive data sets that are difficult to evaluate by eye and impossible to reveal by measurement and simple analysis alone (see chapter 3). Furthermore, these techniques can handle large volumes of data assessable only with difficulty by the unaided mind. (However, whatever advantages accrue through the use of multivariate statistical methods, the technique is still limited by the fact that it operates upon measurement. This deficit we will pursue later.)

Again we may use the example provided by Fisher on the iris. His linear discriminant function (a lineal progenitor of some of the techniques described later in this book), when applied to the known groups of data graphed in figure 3, supplies the discriminating power shown in figure 4. As with the raw measurements, *Iris setosa* is completely differentiable (it could scarcely be otherwise with that data); the overlap of *I. virginica* and *I. versicolor* is now quantitatively defined and is seen to be limited to 13 specimens of the total of 150 (50 for each species) that are examined.

The Characterization and Comparison of Form and Pattern

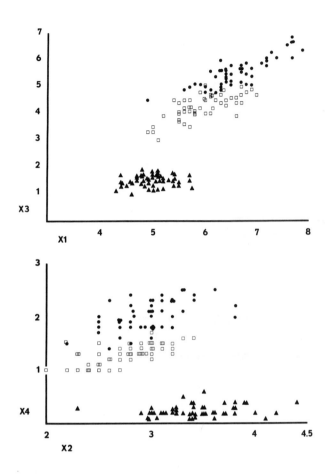

Figure 3. Anderson's *Iris* data. Two bivariate plots of four measurements are taken on each of fifty specimens of *Iris setosa*, *I. virginica*, and *I. versicolor* by Dr. Anderson. Triangles = *I. setosa*, squares = *I. versicolor*, and dots = *I. virginica*. The overlap between *versicolor* and *virginica* is considerable.

Such a study depends upon prior definition of the groups or shapes that we wish to characterize. In some investigations—for example, in those involving assessment of unknown forms such as fossils—we may not know what groups exist in the data; we may not even know if groups are present at all. Under such circumstances a further variety of techniques is available. For instance, grouping or cluster-finding methods may be applied to data already expressed by means of multivariate statistics. Such methods include those presented by Rubin and Friedman (1967), and we have applied them to Anderson's iris data as shown in figures 5 and 6. Here *Iris setosa* is well defined both in terms of the data transformed to a mean of zero and a variance of one, and through the cluster-finding procedure acting in the eigenvector space. But the transformed data do not suggest that the remaining specimens of iris belong to a further two groups; indeed, in figure 5 we are hard put to see three groups in the data. This difficulty notwithstanding, our application of Rubin and Friedman's cluster-finding procedure in the eigenvector space demonstrates all three groups of iris (figure 6), though

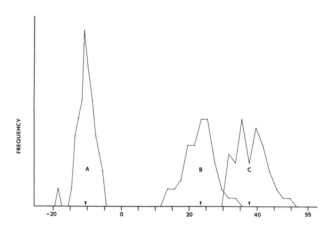

Figure 4. Multivariate discrimination. Fisher's linear discriminant function is applied to the data on *Iris* shown in the previous figure. A = *I. setosa*, B = *I. versicolor*, and C = *I. virginica*. The degree of overlap between *versicolor* and *virginica* is reduced to 13 specimens.

naturally the degree of overlap (18 specimens) is somewhat more than when we already know which is which, as in the use of Fisher's discriminant function. Certainly the group-finding procedure is considerably more powerful than is assessment by eye of the transformed data itself.

It is worth noting that, as a way of analyzing measurements, multivariate statistical methods have their own series of deficiencies. These relate to the theoretical limits (e.g., confinement to normal or near normal distributions) within which data should fall for the techniques to be sound. Accordingly, still other methods are being developed which attempt to analyze data without requiring the imposition of a set of limitations (or at least without requiring the same set of limitations) that are necessary for multivariate statistics. One of these attempts is the neighborhood limited classification of Dr. Peter Neely as described in chapter 5.

Thus the neighborhood limited classification of Anderson's iris data suggests that within the data set are contained the three groups shown in figure 7. One of these groups, *A*, contains all, and only, specimens of *Iris setosa*. Groups *B* and *C*, respectively, contain specimens of *I. versicolor* and *I. virginica*. The power of this group-finding technique is such that only four specimens of these latter two species have been "misclassified" when judged in relation to their neighbors.

It must be emphasized that this is only one attempt. Though the use of neighborhood limited classification may prove ultimately to be a genuine advance, it is also possible that in a few years' time it may be clear that it is a side path in the analysis of data. What is certainly true is that some such techniques are required, and tentative steps should be taken now to discover what they are. To that extent, at least, neighborhood limited classification and other group-finding procedures may prove useful. Again, however, methods of this sort, as with the multivariate analytical approach, are only as good as the measurements that they examine.

9 The Characterization and Comparison of Form and Pattern

Figure 5. Anderson's *Iris* data transformed to a mean of 0 and a variance of 1. Bivariate plots of variable 1 against 3, and 2 against 4. In this case the genera are not identified; it is nevertheless clear from comparison with figure 3 that *Iris setosa* is well demarcated as the smaller group in each plot. The very presence of *Iris versicolor* and *Iris virginica* may be questioned in the larger group, much less the existence of a boundary between them.

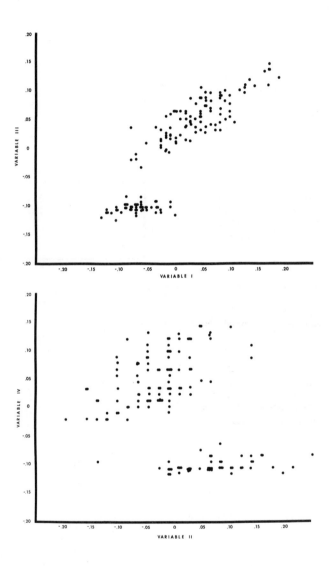

Some Deficiencies of Mensurational Data

This then leads us to look for further deficits in all these studies. And it is immediately clear that the form of the data, measurement itself, suffers from a major fault: measurement provides information only in relation to the points from which measurements are made. There are several problems here.

First, the mensurational data for many studies depend upon particular orientations of specimens. Thus primate skulls are often positioned in the Frankfurt plane. In technical terms such orientations are characterized uniquely (for instance, the Frankfurt plane is defined, as flat surfaces should be, by three points).

Figure 6. Anderson's *Iris* data grouped and plotted in the eigenvector space. In this case the group-finding procedure of Rubin and Friedman has been applied to the data as represented in figure 5, and while *Iris setosa* has been obviously recognized, two other groups that are closer together have also been discerned. These are essentially the two species *virginica* and *versicolor*. There are only eighteen incorrectly identified specimens.

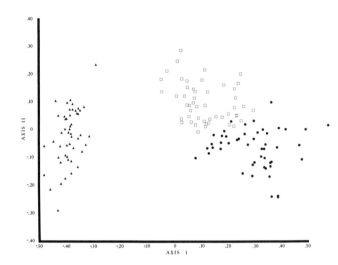

Figure 7. Group-finding procedures. Neely's neighborhood limited classification is applied to the data on *Iris* shown in figure 3. The upper picture is a photograph of the computer output and demonstrates very clearly that three groups exist. The lower figure is a two-dimensional representation of the three groups and shows that *I. setosa* (A) is completely separated from the other two, and that the distinction between *I. versicolor* (B) and *I. virginica* (C) is even clearer than with the cluster analysis of figure 6 or the discriminant function of figure 4. The overlap is confined to two specimens "misclassified" and two outlying to the "wrong" group.

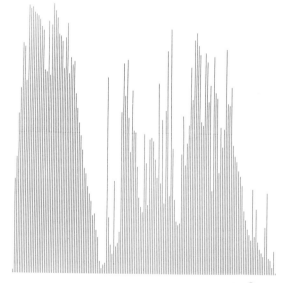

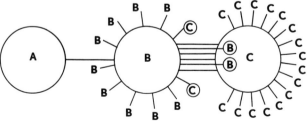

The Characterization and Comparison of Form and Pattern

But the fact is that truly homologous surfaces on different skulls need not be flat, indeed may well be of very different curvature in different skulls; this was probably early recognized as a likelihood but rejected from methodology because of the practical problems involved. More recent attempts to allow for curvature of standard lines and planes are inherent in coordinate measurement.

Coordinate measurements also allow the calculation of any one of a large number of absolute or relative dimensions as required. Such considerations underlie D'Arcy Thompson's use of Cartesian grids and have been recently developed in an example in primate evolution by Sneath (1967), and by Tanner and his colleagues (Johnston, Tanner, Whitehouse, and Sneath, 1967; Tanner, Johnston, Whitehouse, and Sneath, 1969) in the investigation of human growth. This is the technique of trend surface analysis adapted from the work of geologists (e.g., Merriam and Harbaugh, 1964).

Secondly, the question as to the choice of the data points may be made on all sorts of criteria, and these may clearly bias the nature of the information obtained. Although we may make all kinds of clever rationalizations as to how the data points are decided upon, this remains a major subjective element. (It is interesting to note that this fault is also applicable to observation, because it is upon observation that choice of data points depends.)

Thirdly, in order to become quantitative, in order to make measurements at all, we must inevitably reject the information that is available from those points in the shape which have not been chosen as the measurement reference points. Increasing the number of measurements is one of the ways of attempting to avoid this deficit. There are practical limits to this, though it is true that electronic and computer methods are being evolved which can make the taking and recording of data easier so that it becomes possible to obtain extensive data sets that may characterize pattern and shape almost in toto. For instance, computer visualization of radiographs or photographs can allow instantaneous identification of coordinate measurements; flying spot scanners can identify and characterize patterns. But often in such cases the data then become so extensive that analysis, even utilizing the extremely fast computers of the present day, may become difficult (from the viewpoint of programming) and time-consuming, and therefore expensive.

One relatively simple method of characterizing a complex shape has been used by psychologists in tests involving the perception and comparison of nonsense shapes by patients. Figure 8 shows the method which depends upon defining the shape through a variety of angles and lengths that are dependent only upon the positions of turning points in the original shape (Attneave and Arnoult, 1956). Such a characterization looks useful when applied to the convoluted shape shown in the diagram; it appears of much less value when applied to the shape of a bone and it is not easily generalizable to the three-dimensional case. More sophisticated methods are those of Shelman and Hodges (1970), which involve computational techniques applied to data obtained from a picture by scanning (figure 9). Another

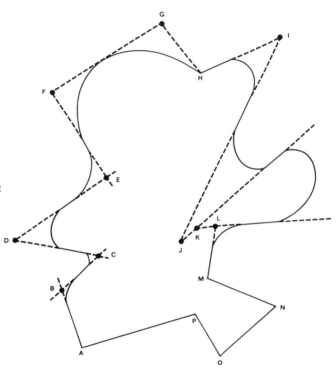

Figure 8. Measuring an irregular contour (after Attneave and Arnoult, 1956). This can be achieved through the use of tangents drawn through turning points in the shape. Though apparently useful for a shape such as that figured here, it is not effective for an irregular object such as the outline of a bone. When applied to a three-dimensional shape, the tangent lines become tangent planes and are excessively difficult to handle.

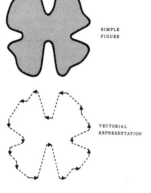

Figure 9. Vectorial representation of shape. The general purpose computer system of Shelman and Hodges is capable of representing complex shapes in a wide variety of ad hoc ways. A vectorial representation is shown above.

method which may turn out to be of considerable value is the medial axis transformation. This has also been developed in relation to problems of pattern recognition (Blum, 1962, 1967) and possesses the attractive property that it represents a simplification of a shape, which may be more easily manipulated but which at the same time contains all the information in the original shape. This method (shown in figure 10 and more fully explained in chapter 8) defines a shape by

Figure 10. The medial axis transformation (after Blum, 1962, 1967). Simple measurements of a shape such as might be taken by anthropologists (e.g., maximum length and breadth—top picture) may be replaced by a medial axis transform obtained by allowing the shape to collapse into itself at a constant velocity in a direction normal to its edge (achieving the medial axis of the middle picture). Medial axes of concave and convex elements (shown in bottom picture) demonstrate that although part of the information about the shape is obtained in the linear medial axis, part is also contained in the velocity of disappearance along the medial axis.

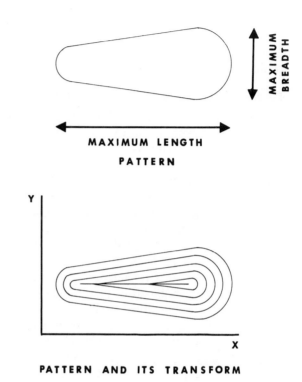

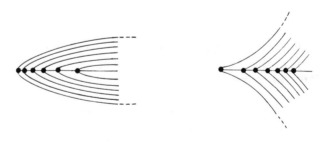

means of a medial axis which is obtained as if the shape had been allowed to disappear into itself with a constant velocity in a direction normal to its boundary. The medial axis thus comprises a curve that can be specified, together with a velocity of disappearance along the curve. By reversing the process one can pass from a medial axis back to the original shape. This technique may be applied to three-dimensional objects as well as flat outlines, though in our hands the programming problems have not yet been entirely solved.

These various questions and answers are leading us to return to using pictorial methods, but utilizing techniques capable of seeing in a picture what is not available to the naked eye.

A Return to the Pictorial Data Set: Optical Methods

It is at this point therefore that we are driven to search further. What is required are methods capable of (a) examining all the information, as is the human eye; (b) assessing it without needing to make subjective decisions; (c) discerning within it problems relating to matters like variation and correlation; and (d) allowing for any kind of data distribution (i.e., for the method to be hypothesis-free). This means a return to the observational view (the picture) for the presentation of data, and it suggests a nondigital (continuous) method of analysis.

It seems possible that optical data analytic techniques may provide at least partial answers. These techniques have been greatly stimulated by research in geology, geography, astronomy, and so on. They depend upon the fact that pictorial data consist of two independent variables (the x and y coordinates) and one dependent variable (the picture density) and may therefore be modeled rather effectively by optical systems which also have two degrees of freedom. Thus for handling pictorial data such techniques are inherently superior to electrical or electronic systems which have only time as the independent variable. They may allow, for instance, the almost instantaneous computation of Fourier transforms as compared with the time for computation that may be necessary (depending upon the size of the data set) for Fast Fourier Transformations on a large computer.

Such techniques are used in a number of branches of science—especially the earth sciences—and the general nature of the results is shown in plate 1. Here Rosenfeld (1969) has utilized the method to characterize optically the complex patterns exhibited by two different types of cloud cover seen during aerial photography. The optical Fourier transform, the technique of which is described more fully in chapter 8, provides a "signature" of the cloud patterns, enabling easy recognition and characterization of otherwise highly enciphered data.

Let it be made clear that these methods are not the whole answer. There are still at least two major deficits in optical data analysis as it is described in these pages. One relates to the fact that some of the information contained in the original picture is still being lost, for one element of the Fourier transform is not as yet capable of being easily recorded and analyzed. Another relates to a limitation of the nature of the picture: it is required to be, at the moment, fairly dichotomous, that is, black and white (it is not that a picture with continuous greys within it cannot give a transform; rather it is that with current understanding, the resulting transform, though it can be used to reproduce the original image, cannot easily provide an analytical result). It is immediately obvious that both of the deficiencies are in a different class from those previously discussed. Although they are genuine, it seems that future studies in physics and mathematics may remove them.

The Characterization and Comparison of Form and Pattern

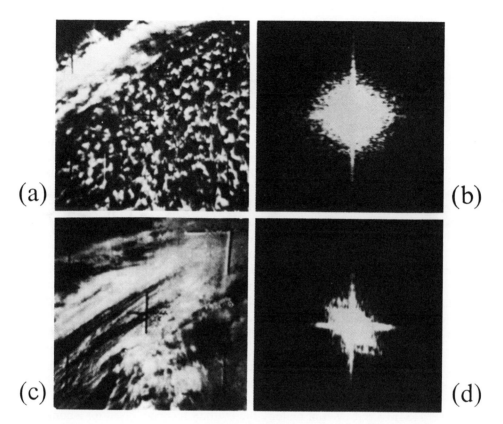

Plate 1. Optical analysis of detailed patterns. Two TIROS cloudcover pictures and their corresponding power spectra. If a picture (a) contains much fine detail and many edges, its spectrum (b) extends relatively far from the origin (central ray). If the edges of a picture (c) have a marked directional bias, the power spectrum (d) departs markedly from radial symmetry.

Current problems in the use of the techniques are thus related to the practicality of the equipment and to the availability of consultative and collaborative help. Nevertheless, twenty years ago many were willing to say that multivariate statistical methods were too complex ever to be used by biologists. Today many biologists utilize them almost as they would a pair of calipers. This may not always be good, for the ease with which these methods are used may lead to their misuse; however, the fact that a technique may be badly used by some should not prevent others from adopting it.

The Functional Role of Form

In all of the foregoing, the problem has been presented as one of shape analysis only. In evolutionary morphology, however, the situation is rendered considerably more complex, because there must be tied into the analyses information relating to the fact that the shapes have functional roles. At all levels (macroscopic, microscopic, ultrastructural) it is readily apparent that whatever complex hereditary and developmental processes result in structure, a considerable element in its formation is associated with impressed mechanical forces. Recent studies of mechano-electric properties of living tissues reinforce and extend that conclusion. Some definition of impressed mechanical forces due to function is therefore required.

This may be obtained through association of the behavior of animals with the relationships of bone and soft tissue structure, especially muscle and connective tissue. Sophistication at this level can be very considerably increased through biomechanical investigations; these may be pursued directly in in vivo situations, using such methods as electromyography, for instance. It may be possible to understand such problems indirectly in terms of the trends that may be revealed in the analysis of shape and form once allometry and correlation can be allowed for by mathematical treatments. It is also possible, however, to test classical anatomical inferences utilizing in vitro techniques of analogy; thus a biomechanical system may be rendered simpler in order to visualize the contained information, though the simplification should not be of so great a degree as to give misleading results. Experimental stress analysis may well be of value in such a situation. In the present studies the physical properties of photoelasticity are utilized. They allow considerable information to be obtained about the mechanical efficiency of biological shapes, especially of the shapes of bones.

These techniques, of course, are not new. Photoelasticity itself has been available since the discovery of the phenomenon early in the nineteenth century (Brewster, 1816). And even in biology the method has been used for a considerable period for elucidating the load-bearing efficiency of different anatomical structures (Le Gros Clark, 1945; Pauwels, 1948). The example of Pauwels (1965; diagrammed in figure 11) demonstrates vividly the load-reducing action of certain of the tension-bearing structures upon the skeleton of the upper arm. The technique (more fully described in chapter 6) is being used here to investigate differential mechanical efficiency of the shapes of primate bones in relation to their varying functions within the life histories of the animals.

The Time Dimension: Fossils

It is at this point in such a series of studies that problems of a different nature are encountered. In the basic researches as they have just been outlined, materials from extant animals are available. For living forms, clues are provided by (a) other facets of the life processes of the individual and of the species, (b) the researcher's ability to examine numbers of specimens, and (c) the relative completeness of data sets. These result in conclusions having reasonable degrees of likelihood. But further stages in such studies can bring to bear on the conclusions the unique property of fossil data: that account may be taken of time and continuity. There then result the difficulties of fitting, into well understood data, information from single specimens—information that may well be incomplete and that may also

The Characterization and Comparison of Form and Pattern

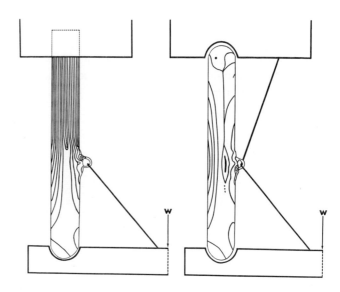

Figure 11. Photoelastic analysis of a simulation of an elbow. The vertical element represents the humerus, and the horizontal bar with its load represents the radius and ulna. When a single tension member (representing brachialis muscle) is included, then the load on the lower half of the humerus is relieved (the numbers of dark lines are fewer than in the upper part of the model). When a second tension member is added (representing coracobrachialis muscle), then all of the humerus is relieved of the major part of its load. Thus it is possible to demonstrate the stress-relieving effects which the particular muscular disposition has on the bones.

involve a certain degree of reconstruction. Questions relating to single specimens and incompleteness of data are statistical matters which are reasonably clearly known, if not completely understood. Problems relating to reconstruction are more difficult to allow for because there is always the chance, sometimes quite high, that the investigator will unconsciously channel the nature of the reconstruction in terms of previous conceptions about a fossil.

Some attempts have been made in these researches to study the incorporation of data from unknown forms. In the main this has meant working with published dimensions, and, with great circumspection, working from casts or even photographs of specimens. Such methods are required because of the difficulties of obtaining data directly from originals; it means that any conclusions derived therefrom can be considered only as pilot or tentative in nature. But it has also been possible to treat some extant forms as "unknown" test pieces. Such strategies allow considerable investigation of the biological problems relating to interpolation. Interpolation has been investigated through the mathematical methods of data handling and also in terms of the physical techniques of experimental stress analysis (chapter 7).

These then are the purposes of the present work: to review current morphological studies of primates by utilizing mathematical and physical ways of observing data; and to assess the extent to which they provide new insights into problems of characterization of shape and form, especially within the framework of evolutionary adaptation to biological function. The applicability of the methods in related fields of biology may also be of interest.

Before these purposes can be achieved, it is necessary to have a basic understanding of functional morphology in the anatomical regions chosen for study. Thus, the next chapter reviews some general aspects of the functional-structural equation and outlines particular applications in the study of the shoulder, hand, and hip of the primates.

2 Associations between Function and Morphology

Introduction

However careful the techniques of data collection and however detailed and complicated the methods of analysis, new insights into morphology cannot be expected if we do not take into account the fact that biological shapes exist within the limits of biological functions. Accordingly, in initial studies of shape, some information is needed about those aspects of the behavior of animals that have implications for the anatomical shapes concerned.

At one extreme, it might be thought that we require a relatively complete recognition of the detailed behaviors of which animals are capable. With the present status of research, this large gap in our knowledge is gradually being filled, especially by workers utilizing new hypotheses, techniques, and methods of analysis. However, for the great majority of genera, it will be some time, if ever, before all information can be garnered.

For instance, exhaustive study of the total locomotor pattern of a primate troop pertains to ecology and social behavior. It includes a whole range of data about environmental variables such as the topography of the general locality, the more detailed architectural features of the microniche (e.g., the shape, surface features, dendritic patterns, and strengths of tree branches), and about temporal changes such as seasonal flooding or foliage succession in vegetation. Direct information about locomotion should undoubtedly include feeding, resting, and sleeping postures; varieties of gaits; the duration, size, and frequency of leaps; the contexts in which alternative structures for locomotor patterns are used or avoided; the ontogeny of locomotor development; and so on. In all these and still other parameters, differences between troops should be estimated from the range of the whole species. Clearly the above series of studies could require a lifetime's commitment to even a single group. A reminder of the total number of currently recognized extant primate genera—59 according to Simpson (1945)—suggests the virtual impossibility of providing such data for evolutionary comparisons throughout the order.

At the other extreme, however, it is obviously inadequate to utilize concepts of function within behavior that relate to such broad and (especially within the primates) almost meaningless terms as "terrestrial" and "arboreal" when applied to entire locomotor modes.

Given the impossibility of obtaining the former detailed answer and the overly simplistic nature of the latter generalization, can we define practicable intermediates? What elements of the behavioral structure as it is now known are of most relevance in the understanding of morphology? The answers to such questions are likely to differ according to the nature of the behaviors, the particular animals, and the anatomical regions involved.

A Strategy of Investigation

Thus, for living forms, lines of argument can be followed from behavior in the field to detailed morphology in the laboratory. In considering parts of the postcranial skeleton, major steps in such an argument are those shown in figure 12.

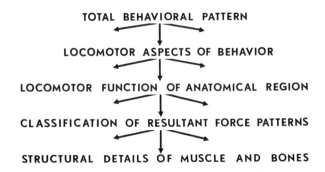

Figure 12. Relationships between function and structure. This attempts to show the multiple levels that may be considered in the theoretical arguments, and the one-to-many relationships that presumably exist at each level.

As we have just seen, the total behavioral pattern can be defined for scarcely more than a few primate species. Nevertheless, only a relatively small amount of information about behavior is necessary to supply sufficient data for the next step of the argument: the *locomotor aspects of behavior*. This information can be obtained from those field studies that have been concerned with locomotion (e.g., parts of Carpenter, 1940); from a number more in the field of the natural history of the primates; and from a large number of anecdotal reports which, originating independently, provide additional facts. From these studies concerning primates in their natural surroundings and from laboratory researches of locomotion in living forms (such as the studies of Avis, 1962; Hall-Craggs, 1965; Hildebrand, 1967; Tuttle, 1967), simple deductions can be made about the *locomotor functions of particular anatomical regions*. (The information that is lacking when all aspects of behavior are not considered is relatively unimportant in this context as long as the behaviors which are missing do not produce large forces, as we shall discuss below).

Having defined locomotor functions, we may then obtain some concept of the forces operating in particular anatomical regions. Here again, a given function may produce a rather large number of complexly arranged forces which cannot be defined in toto. But once more, complete definition is not required; those forces which are small may be neglected, while those which are large will be more obvious and can be more readily distinguished. The *classification of these force patterns* may proceed by means of analytic studies of the mechanics of complex locomotor movements, and the end results may be relatively uncomplicated. In addition, the characterization of forces may be studied synthetically by experimental methods which start with a simple analogy and gradually increase in complexity.

It is at this step that correlations may be sought between, on the one hand, the derived forces acting upon the part and, on the other, the *detailed muscular and bony structure of the region*. This may be done in two stages by examining differences first in muscular size, shape, and disposition and secondly in bony morphology, both shape and architecture.

Because the material in most studies utilizing soft tissues tends to be limited, it is often only feasible to investigate muscular relationships at a crude level.

Nevertheless, in his extensive studies of the primate hand and foot, Tuttle (1969) utilizes long series of wet materials and is thus able to group his data at specific and even subspecific levels.

Owing to the greater speed of osteological work, to the greater abundance of materials, and to the quantitative nature of many of the results, it may be possible to find bony correlations not only with the initial crude functional classifications as utilized for soft tissues, but also with the finer points of more detailed force patterns. Indeed, it may well be possible to reject cruder groupings which have nevertheless been helpful in earlier studies.[1]

Within an apparently linear argument, there are steps which require further elucidation. One such step relates to the need to know only those elements of behavior which produce relatively large forces in a given anatomical region (e.g., compressive forces in the shoulder caused by landing after a leap), rather than other aspects of behavior which may be vitally important in the life of the individual but which produce rather small forces in the given anatomical region (e.g., forces in the maternal shoulder related to cradling a suckling infant). A second similar step concerns locomotor functions; again we may base our classification of force patterns upon locomotor functions of the particular anatomical region rather than upon a classification of the locomotion of the whole animal (if indeed it is feasible to attempt the latter).

Certain points need to be emphasized in examining these particular steps in more detail. It has long been established that one element of the plastic response of bone is to some average measure or resultant of forces acting upon it over a period of time, rather than to individual forces themselves (e.g., Murray, 1936; Washburn, 1947). Recent work on piezoelectric and semiconducting phenomena within bone suggests more precisely the mechanisms by which this is achieved (Rothmann, 1967). If selection for mechanical efficiency is occurring within populations of individuals, then the inheritance of bony morphology will be related to, among other things, a resultant force related to the mechanical demands of

---

1. In earlier studies (e.g., Oxnard, 1963; Ashton and Oxnard, 1963) in which it was not possible to provide analyses that could survey all of the data at one step, investigations were simplified by examining the many genera as a small number of groups defined by the function of the shoulder in locomotion. It is clearly much easier to study variation in a number of muscular and bony variables if comparisons have to be made among only a small number of groups. Such a crude grouping or classification cannot be applied to the locomotion of the entire animals (although this is how it has been interpreted by some authors, e.g., B. G. Campbell, 1966).

In later studies (Ashton, Healy, Oxnard, and Spence, 1965; Oxnard, 1967, for example) the multivariate analyses used are perfectly capable of dealing with large numbers of variables, specimens, and groups. Accordingly, in these studies coarse groupings are not utilized; the multivariate statistics operate on individual genera, a finer level of detail which is generally agreed upon by most primatologists. It is of some considerable interest therefore, that these later studies (especially Oxnard and Neely, 1969, and Ashton, Flinn, Oxnard, and Spence, 1971) demonstrate a coarse grouping of genera, similar, though not identical, to the initial grouping based upon consideration of the function of the shoulder in locomotion. This grouping, obtained in 1966 and later, is not, of course, based upon functional information; it is purely morphological. The fact that it corresponds with the functional (locomotor) grouping noticed several years before (Napier, 1961; Oxnard, 1961), and that it was clearly foreshadowed by a number of even earlier writers (Napier, 1959; Napier and Davis, 1959; Erikson, 1954; B. Campbell, 1937) is of extreme interest.

function.[2] It is therefore necessary to attempt to derive this resultant; pending more elaborate studies, this is supplied by a classification of force patterns based upon functions producing major forces.

Another reason why, in studying extant forms, it is necessary to attempt to use simple classifications of resultant forces acting upon individual anatomical regions rather than classification of the entire locomotion of the animals, is that this information must also be of assistance in the study of fossils. Thus in the absence of parallelism of overall function in different forms, it may be difficult to decide (unless there are obvious mechanical correlates) whether a particular structure is found in one animal because it moves in a particular way or because it belongs to a particular taxon. (It is, of course, not unlikely that both reasons might apply.) The use of a classification of functions operative in individual anatomical regions pinpoints common factors in locomotion, factors which may well be obscured by the consideration of the overall locomotion (or more widely, the behavior) of the whole animal. For example, the information from certain Prosimii has proved of value in testing hypotheses about relationships between bony shape and locomotion in apes. Thus the common factor in the function of the shoulder in the potto and the gibbon is that the region is subject to tensile forces. In these two animals the tensile forces do not act in the same way; in the gibbon the body is most often suspended vertically from the forelimb alone, and the movements of this unit are fast and furious; in the potto the body is sometimes suspended from the forelimb with the hindlimb taking a proportionate share of the body weight, and on the whole the movements of this animal are the reverse of speedy. Yet there are many facets of muscle and bone morphology that are similar in the two forms that are not shared by other forms closely related to each, and that appear mechanically relevant to the concept of the shoulder region being adapted to tensile stresses. Thus in a behavioral situation where one or several biological functions occasion large forces and most other functions produce relatively small forces, then the resultant will be related more to the former than the latter.

If, therefore, we can identify elements of behavior that produce relatively large biomechanical forces and that are therefore major contributors to the resultant, these elements will have a discernable effect upon the adaptive shape of bones. This is likely to be the case in, shall we say, the forelimb in animals which utilize that organ for locomotion. Thus, however much a gibbon may use its forelimb in feeding or grooming, by far the greatest forces acting upon the *shoulder* will be those related to its function as the highly mobile tension member that joins the gibbon's body to its limb during various kinds of under-branch movement. Likewise, however much a gorilla may use its forelimb for manipulation

---

2. A more elaborate exposition of these ideas is presented by Stern and Oxnard (1973) in which the concept of "resultant" as used here is further discussed. As this anatomical resultant is only partly equivalent to a resultant in mechanics, and especially because the time during which a force acts is also included in the notion of the anatomical resultant as used here, Stern and Oxnard utilize the more accurately descriptive term "anatomical momentum."

of small objects, by far the greatest forces impinging upon its *hand* are presumably those produced by the hand's use as a propulsive strut in knuckle-walking, in which a large part of the animal's appreciable body weight is supported and impelled by that organ. It is to these locomotor aspects of behavior, and, within locomotion, to those individual movements producing the greatest forces, that one may expect to see the most clearcut morphological adaptation.

It is nevertheless possible that some of the remaining functions of a given anatomical region (the gibbon's shoulder, for instance) may involve forces that are large enough to have a discernible effect upon the resultant. This is especially true in the case of the gorilla's hand, the remaining functions of which, such as reaching for and pulling towards itself food or nesting materials, and reaching and pulling its own body upwards in climbing (more frequent in young animals), may well summate to a function of the hand as a gripping hook. Such a summation may be expected to have a considerable effect upon the resultant force acting upon the hand of the gorilla; evidence of this may also be expected in its skeleton (figure 13).

In total contrast, the forelimb of modern man performs a very large number of different functions, none of which is powerful enough to contribute more markedly than others to the resultant force acting upon the forelimb. Of these various functions, perhaps those most liable to have effects upon the resultant in the shoulder region are those related to (a) the shoulder acting as a rotatory and propulsive agent, as in spear or rock throwing (important, no doubt, at early stages in the evolution of man), or (b) the shoulder acting as a tensile member in the lowered position when heavy weights are being supported by the dependent arm (as in carrying a shopping bag in a more recent evolutionary period). For the human

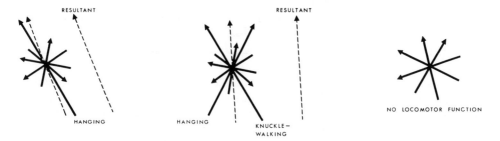

Figure 13. Forces acting upon elements of the forelimb. Left diagram: forces acting upon the element in those animals (e.g., the orangutan) in which the primary locomotor movement is hanging. The resultant force is close to the main force due to hanging. Other forces (e.g., due to manipulation) are small and contribute little to the resultant. Middle diagram: forces acting upon an element of the forelimb in those animals (e.g., the gorilla) in which there are two locomotor movements, one associated with knuckle-walking, the other with hanging. The resultant force is determined primarily by these two forces. Again, other forces are small and contribute little to the resultant. Right diagram: forces acting upon an element of the forelimb in man where locomotion has been removed as a function. There are no large forces and the resultant is consequently differently oriented and rather small. Presumably evolution of the forelimb in man has proceeded from force situations like the first or second to the third.

hand it is possible that, of the many functions of which it is capable, the one most likely to have the biggest effect upon the resultant force may be the power grip employed in numerous manual functions.

In man, however, the matter is by no means so clear cut as with nonhuman primates. The personal predilections of the researcher may well suggest for emphasis other aspects of shoulder and hand function; until appropriate in vivo biomechanical studies are performed, there will be little objective reason for choosing one explanation over another. In any case, even biomechanical studies are unlikely to settle the matter. Occupational, recreational, social, and probably many other differences in modern man may well interfere with assessments. Accordingly, it is improbable that in the human hand or shoulder we will see any particular adaptations to particular functions. More likely, in both anatomical regions there will be overall adaptation to resultants not necessarily similar to any specific subset of forces (other than by chance). Presumably adaptations will then be to shoulder mobility and manual dexterity which are more extreme than in any of the other primates but which also lack the power that characterizes the majority of nonhuman primate forms.

Indeed, perhaps the most significant difference in man as compared with the previous examples in nonhuman primates is the absence of frequent functions producing large forces in the shoulder and hand (figure 13). We can draw an association with this in the macroscopic external configuration of the scapula; that of man possesses thinner bony plates and smaller and smoother buttresses for its size as compared with those of the comparably sized gorilla and the small-bodied gibbon. Similar observations may be made at an internal architectural level in the arrangement of compacta and trabeculae in the hand bones of these examples. In man there is little compact bone; most of the architecture consists of finely carved trabeculae. In the great apes, in contrast, the phalanges contain only a little trabecular bone; the major part of these elements consists of heavy compact lamellae.

Thus it can be seen that if it is possible to pick out a small number of functions within behavior as producing large forces, then it may be somewhat easier to outline bony adaptations; in general, this is the case for postcranial remains of nonhuman primates in which all limbs are used in one way or another within a locomotor context. If it is not possible to identify a subset of functions that contribute largely to force patterns, then it may be much more difficult to define adaptation of bone. This is the case for some regions in man. It may sometimes be possible to identify the periods during the evolution of man when the changeovers occurred—that is, when a small number of obvious functional parameters ceased to be of importance in appropriate morphologies (figure 13). This is likely to be the case in studies of the evolution of form and function in certain anatomical regions such as the forelimb, face, and jaws. But it is not so in studies of the evolution of, say, the pelvis and foot. For in man, in these latter regions, we do not have the example of a set of large forces replaced with a set of small forces; rather we have a changeover to a different set of large forces associated with walking and running on two legs (figure 14).

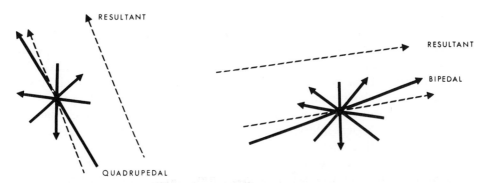

Figure 14. Forces acting upon elements of the hindlimb. Left diagram: forces acting upon the element in those animals whose main mode of movement is quadrupedal. As in the previous figure, the forces due to locomotion will be large (large arrow) and hence will most resemble the resultant. Forces due to other functions, such as prehensile or manipulative functions of the foot, will be small and have relatively little effect upon the resultant. Right diagram: forces acting upon the element in man where the principal mode of locomotion is bipedal. Here the forces due to bipedality will also be large, but oriented totally differently from the principal force in the left diagram. The resultant will be likewise totally different from that at the left. Again, other small forces will be present that will make but little contribution to the resultant. In this case, in contradistinction to the situation modeled in figure 13, the evolution of the hindlimb must have proceeded in relation to a change from a force situation like the left to one like the right.

This theoretical discussion can be taken a number of steps further because the relationship between function and structure is so complex and by no means clearly understood. For instance, we can see that a particular structure may have as its basis within an individual a whole series of "causative relationships." Even though in the end these must be drawn together so as to produce a single, structurally harmonious element, it is nevertheless possible to see that there may be, in theory at any rate, many (probably overlapping) morphogenetic channels. For the structural features of bones (one end of the functional-structural axis under consideration here), these morphogenetic effects certainly include, among others, the effects of stress bearing, genetic constraints, blood supply, and growth. Although these are all melded in the final product through ontogenetic mediation, it is possible to isolate principal determinants to some degree or other, either through experiment or through morphological analysis.

At the other end of this functional-structural axis are the many different facets of behavior. These too have their causative effects upon the structure of the individual through, among other things, the mediation of statistical reproductive survival mechanisms (Stern, 1970). Some mechanisms may act through functions that occur only rarely but which, when they do occur, are determinants of life or death. Others may act through functions that confer only slight advantages in a wide variety of ways. The functions involved may relate to behaviors as different as communication, predation, defense, and so on, interrelated though these may be.

The situation is thus one in which we might say that there is no simple one-to-one relationship between structure and behavior. A given structure may be compatible or even important for a number of behaviors. A given behavior may be more difficult or even impossible without many different structures. In addition — and this is one of the elements that further clouds the issue — it is always easy to suggest associations or correspondences between structure and behavior in superficial terms. How often have we, as morphologists, been able to glibly hypothesize a particular relationship between structure and function on the basis of some newly discovered morphological fact? How often have we also known that if the morphological information had happened to be exactly the reverse, a superficially attractive relationship could still have been postulated?

Thus, though a large number of structural-functional relationships may be noted, many of these may be scientifically spurious, in the sense of an accidental association or correspondence. In other words, the interface between structural and behavioral plasticities may be rather small, being in the nature of a bottleneck or filter. One of the tasks of functional morphology relates to the elimination of those suggestions that have no scientific validity. This process may be carried out by detailed and careful research designed to test the ideas; however, final reliance on anecdotal "proofs" of such relationships can only obscure the matter (though anecdote may well be one of the initial stimulating factors suggesting hypotheses for testing, as also may previous knowledge, reason, or dreams).

How in fact can the reduction from possible to probable structural-functional relationships be achieved? One technique has already been suggested: the careful examination of individual associations in terms of the more detailed levels of association that should exist within each. This reveals clearly that any given vertical argument passes in reality through a series of horizontal steps (see figure 12). Each of these steps can be studied, some more easily than others.

Morphological Studies of the Primate Shoulder

First of all we may look at the shoulder region, which has been examined through the comparative myological and osteological investigations of Ashton and Oxnard (1963, 1964). They have confirmed and extended information available from a number of studies pursued by various workers over many years (e.g., Frey, 1923; Miller, 1932; Inman, Saunders, and Abbott, 1944). They demonstrate the close interrelationships that exist between functions of the shoulder and its structure, and they form a necessary background to studies presented in later chapters.

In the first instance (Ashton and Oxnard, 1963, 1964) a series of dissections on a wide range of primate genera presents a full comparative spectrum. They reveal a number of meristic characteristics of the shoulder muscles (for example, whether the atlanto-scapularis anterior muscle lies superficial or deep to the trapezius muscle); such findings comprise a large part of the myological literature of the previous century. But the main variations to which the study draws attention are those affecting muscular size, form, and orientation. The anatomical peculiarity of these latter contrasts lies in the fact that many cut across the current taxonomic subdivisions and correlate well with parallels of locomotor function in the shoulder in the different species. The mechanical significance of these differences cannot be assessed critically until adequate experimental studies have been undertaken (e.g., Basmajian, 1972). But careful anatomical inference indicates many functional relationships that are mechanically meaningful for locomotion.

One part of the locomotor movement of the shoulder is the protraction of the forelimb with the hand free as a preparation for the propulsive stroke. In terrestrial species such as baboons this is a less powerful and less extensive movement; the entire forelimb may be smaller; it is moved forward to a lesser extent, and the movement is aided considerably by gravity. In the more arboreal forms, in contrast, the forelimb, which is likely to be a bigger, heavier object, may have to be moved further forward, or even above the head against the action of gravity; in addition, the mediolateral mobility of this movement is likely to be considerably greater because of the three-dimensional nature of the sub-, pre-, or super-stratum. These requirements are presumably less in creatures which more often run quadrupedally on large branches, (e.g., squirrel monkeys and guenons) and are increasingly greater in species which more frequently live in small branch environments (e.g., woolly monkeys and proboscis monkeys). The biomechanical constraints must be greatest of all in those capable of hanging or swinging under branches (e.g., spider monkeys and gibbons). Most of these differences are additive; that is, animals which often raise the forelimb above the head do this as a locomotor movement which is additional to those movements of species which only sometimes raise the limb above the head. It is only in the very highly adapted lesser apes (and perhaps the spider monkeys) that the extent of arm-raising activities is so developed as to obscure the basic quadrupedal nature of locomotion.

Again, correlating well with these functional differences, there are found more extensive developments of the arm-raising muscles in the arboreal forms than in their more terrestrial relatives. This is especially so in those tree-living species which frequently move and forage in a small branch environment, and to the greatest degree of all in those which truly brachiate (figure 15). Thus the deltoid muscle, the upper part of the trapezius, and the lowermost part of the serratus magnus are all highly developed in the more acrobatic species. In addition, the direction of the fibers of the different muscles is more efficiently placed for arm raising. The cranial fibers of the trapezius muscle are directed further laterally so that they are efficient rotators of the scapula rather than simple protractors. Together with the increased number of heavier caudal digitations of the serratus magnus, these fibers help to form a more efficient force couple for scapular rotation during the raising of the arm in front of or above the head. The bigger and more complex deltoid muscle is inserted farther down the arm, presumably increasing its efficiency as an abductor of the humerus.

During quadrupedal locomotion a second movement is the propulsion forward of the body by retraction of the forelimbs (in addition to the action of the hindlimbs), the hands being fixed on the locomotor surface. Though in most terrestrial

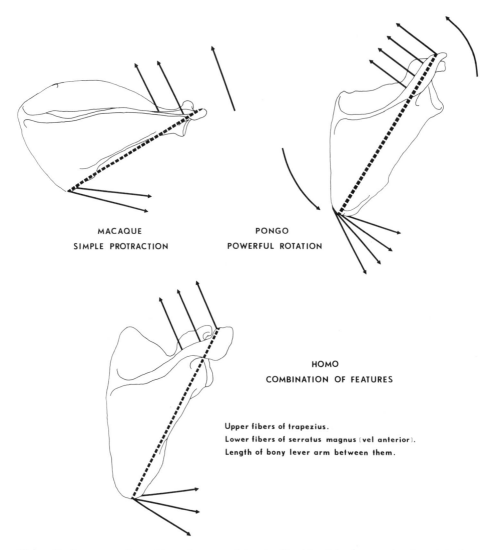

Figure 15. Some muscular and bony features of the shoulder. Top left: the scapula of a relatively quadrupedal type of primate (*Macaca*) in which part of the muscular mechanism for producing scapular rotation is figured. Two of the muscles are shown (the cranial part of trapezius and the caudal digitations of serratus magnus). They are relatively small (as demonstrated by being composed of only two arrows) and their alignment is such that they are not particularly effective in rotation because they are disposed nearly at right angles. It would appear that they are mechanically most efficient for scapular protraction and stabilization of the medial border of the scapula.

Top right: the scapula of relatively acrobatic orangutan (*Pongo*) in which the equivalent part of the muscular mechanism is shown. Here the muscles are much more powerful (composed diagrammatically of four arrows each) and they are more nearly parallel, hence having a considerably greater effect on the rotation of the bone that is necessary in arm-raising in an acrobatic animal.

The bony feature that appears to be correlated with these muscular features is shown by the dotted line. It is the relative (the diagrams are approximately the same scale) length of the lever arm between these muscles. It is short in the form (*Macaca*) where scapular rotation is less important; it is long in the acrobatic orangutan. The bottom illustration shows, for comparison, these particular features in man.

monkeys (e.g., baboons) the action of the forelimbs is likely to be less powerful than that of the hindlimbs (which are presumably the main propulsive agents), the forelimbs become proportionately more important among arboreal monkeys. While running across a broad tree limb (an activity characteristic of squirrel monkeys and guenons, for example) is not so very different from running on the ground, climbing activities in a small branch environment and acrobatic activities characteristic of many monkeys (e.g., howler monkeys, woolly monkeys, proboscis monkeys, and guerezas) are dependent upon an increasing importance of the forelimb. The epitome of this is reached in those animals which habitually arm-swing when they move (e.g., spider monkeys and gibbons). Especially in the gibbons, the hindlimbs scarcely participate at all in propulsion; almost the whole body weight is propelled forward by retraction of the forelimb.

Associated with these locomotor differences are differences in the relative mass of the propulsive muscles; in arboreal primates the relative mass shows considerable increase over that in primates which are mainly terrestrial. The greatest increases in relative mass of propulsive muscle are found in monkeys that are the most acrobatic; the extremes are the gibbons and siamangs. Although the free-ranging gorilla almost never moves by arm-swinging, and the chimpanzee does so only rarely, both species are capable of arm suspension and arm-swinging in a way denied to almost all monkeys; the young of both are particularly good at acrobatic and arm-swinging activities. The findings of relative muscular size and disposition in these species appear to be associated to some extent with these abilities,[3] because biomechanical aspects of the musculature of both the free-ranging gorilla and the chimpanzee resemble somewhat those of the highly acrobatic gibbon (fast) and the orangutan (slow).

These contrasts in relative muscular mass are especially obvious in the latissimus dorsi muscle and seem to be related to its greater mechanical advantage for retraction of the forelimb in species for which retraction is proportionately more important in locomotion. But other structural parameters of the muscle — for instance, the more vertical direction of its fibers and its increased ventrolateral extent — appear also to be related mechanically to these concepts. Similar, if somewhat less marked, changes are present in other muscles which aid in propulsion (for example, the entire pectoral mass).

In addition to these rather precise morphological adaptations, which apparently are related to individual facets of the locomotor movements, there is also a series of general adaptations. These general adaptations seem to be related to an increasing tensile force that is placed upon the shoulder as a whole during acrobatic activities. It is clear that this force must be a considerable portion of any type of movement that can be designated arm-swinging, as in the greater and lesser

---

3. A considerable problem exists here. Because it is highly likely that in times past the chimpanzee and gorilla as adults were considerably more arboreal and acrobatic than at the present day, because at the present time the immature forms of both species are indeed more arboreal and acrobatic than are the adults, and because *some* of the morphological features adaptive to an acrobatic arboreal existence may also be adaptive to the peculiar terrestrial locomotion shown uniquely by these forms, it may be especially difficult to separate these different associations. Undoubtedly the relationship between function and morphology may exhibit a certain degree of time lag.

apes; it is also involved, though to a lesser degree, in the locomotor movements of the arm within the small-branch environment, as in the woolly and proboscis monkeys; to an even lesser degree it is an element in quadrupedal activities associated with running on relatively large branches in the main canopy of the forest (squirrel monkeys and guenons); and it is presumably of minimal significance in truly ground-living creatures like patas monkeys.

Among the morphological features that appear to be related to this tensile force are the powerfully developed ventral fibers of the latissimus dorsi muscle which, when the arm is raised, lie almost vertically and help to transmit the body weight from the trunk to the arms. Again, in this position, the thick caudal digitations of serratus magnus are almost vertical and transmit the weight from the lower part of the trunk to the scapula. Finally, the powerfully developed deltoid muscle lies vertically during suspension and, along with the short scapular muscles, helps to transmit weight from the shoulder girdle to the humerus. Though these factors are greatest during body suspension, they will also be related to pulling the body upwards, as in nearly vertical climbing, when the hind limbs may aid by pushing from below. The more that the movements of locomotion involve tensile forces acting on the whole region, the more these features of the shoulder musculature are developed. For instance, these particular features are very highly developed in the potto and angwantibo, animals that do not move in any way resembling that of the various monkeys and apes save that during the under-branch progression of which they are capable the shoulder is subject to tensile forces.

The detailed myological adaptations revealed by this study point to features of the bones that may be associated directly with them (Ashton and Oxnard, 1964). These do not comprise such features as the supraspinous fossa index, a standard anthropological measure. Instead they include such bony characters as the position and orientation of muscular attachments, the relative lengths of muscular leverages, and the positions and orientations of joint surfaces. Measurements are devised to define, for example, (a) the relative distance from acromion to inferior angle, giving information about the couple arm between the upper part of the trapezius and the lower digitations of serratus magnus; (b) the angle at which the glenoid cavity is placed on the axillary border (a feature associated with the upward orientation of the shoulder joint); and (c) the relative clavicular length, which reflects the extent to which the shoulder joint is placed laterally on the body wall.

The independent examination of a total of nine such osteological features supplements that of the muscles. Many show parallel variations in different species which may be related to the functional differences just outlined. The nature of the differences, furthermore, is often such as to increase the mechanical advantage of the muscular variations that have already been observed. The bony lever arm between the trapezius and serratus magnus in acrobatic species is long (figure 15), thus correlating with the increased efficiency of the muscular mechanism for scapular rotation provided by the greater development of these two muscles. Again, apparently in relation to the more upward movements of the humerus in acrobatic species, the glenoid cavity is oriented more cranially; the more acrobatic the animal, the more this is so. The shoulder projects the most laterally, thus giving more freedom to the shoulder joint, in those forms whose environment is most obviously three-dimensional—especially in brachiation proper, but also in moving in the small-branch environment and to some degree in all climbing activities. Detailed inspection has shown that in all the bony dimensions there are gradations among the different primates that parallel the extent to which they are acrobatic within the trees.

It is clear, however, that in the examination of features individually, the information given by each is not additive. The more statistically intercorrelated are the individual features, the less new information about the shape of the shoulder is provided by each. If, for instance, any two of the dimensions are fully correlated, each alone is the equivalent of both together. Accordingly, the next stage of such investigations includes the multivariate statistical analysis of the data. We shall consider these problems in the next chapter.

Another Example: The Primate Hand

The studies of Dr. Tuttle on the hands of hominoids show evidence of the same detailed series of behavioral and morphological associations (Tuttle, 1967; 1969). Because these investigations also form a considerable part of the background to studies presented in later sections of this book, Tuttle's work will be summarized here. He has shown that the hands of the Hominoidea evidence four adaptive modes which distinguish the lesser apes, the orangutan, the African apes, and man.

The hands of all of these creatures are adaptive for manipulatory functions that fit well with the inquisitive nature of higher primate behavior, though it is clear that a large part of this adaptation is to be found in the brain rather than in the hand. The morphological correlates of manipulation are chiefly in the thumb, in contradistinction to the view of many authors who have tended to envision the hands of apes as being simply hooks for arm-swinging or inverted hooks for knuckle-walking. Tuttle has demonstrated that though the thumbs in these animals are reduced relative to elongated fingers, they are certainly far from being functionless and are most important in fine movements.

However, the major features of the hand of the apes, especially in terms of forearm and finger elements, are undoubtedly related to locomotor compromises. The hands of all four apes are adapted to arboreal movement, foraging, and feeding in terms of grasping with power. Superimposed upon this basic arboreal adaptation are requirements for richochetal brachiation in the small-bodied gibbons, slow acrobatic hanging and grasping in the heavy-bodied orangutan, and knuckle-walking in the chimpanzee (carried to an extreme in the gorilla).

Thus in the gibbon is seen a series of morphological features with emphasis on flexion. For instance, the total relative mass of the flexor muscles of the forearm are the greatest in the hylobatines. Tuttle has shown that this is almost entirely due to digital flexors as opposed to wrist flexors (which are in fact very similar in all hominoids). There are also marked differences between the superficial and deep flexors, and these correlate well with the biomechanical situation.

The intrinsic muscles are positioned so as to have greater mechanical advantage in flexion.

Associated with these muscular features of the hylobatine forearm and hand are many osteological features that are equally sensible when seen in the light of the functional situation. Thus the hooklike appearance of the hand may be attributed not only to the volar curvature of individual elements of the digital ray, but also to the flexed position of many of the joint surfaces.

Similar associations between morphology and locomotion are found in the great apes. Thus in the orangutan there is also heavy dependence upon flexor musculature, although not in the specialized way that is found in the gibbon. Powerful flexor muscles are capable of supporting these heavy animals for considerable periods of time in suspended postures. As Tuttle has pointed out, some of the differences are undoubtedly due to the fact that in richochetal brachiation (e.g., gibbons) there is a requirement for the rapid release of the substrate; whereas in slow swinging and climbing (e.g., the orangutan) other requirements obtrude themselves: for example, the free dorsi-flexion and adduction of the wrist and the independent extension of the fingers that are of special advantage in slow climbing in peripheral foliage when an animal must frequently draw in enough small twigs to support its weight. This allows the animal to hold some twiglets while remaining digits seek for additional support. If some twigs break, the animal does not have to open the whole hand in order to grasp replacements. Although chimpanzees and gorillas share some of these features with orangutans (figure 16), and though they have remained arboreal as infants and in foraging, feeding, and nesting behaviors, the African apes have developed a number of functional morphological adaptations to the demands of terrestrial knuckle-walking locomotion (figure 17), as Tuttle has shown.

The major morphological features that seem to be related to the African apes' assuming knuckle-walking postures and movement are associated with maintaining the integrity of the wrist and the second to fifth metacarpophalangeal joints. Thus extreme dorsal, medial, and lateral displacements of the wrist are not possible in these species as compared with the Asiatic apes. The notable shortening of the long digital flexor muscles in African apes is probably a special adaptation for maintaining the metacarpophalangeal joints of the second to fifth digits in the position of hyperextension with minimal muscle activity during resting stances. In locomotion, as the load moves over a supporting hand, the digital flexor muscles are probably placed in a state of stretch. This presumably greatly facilitates the propulsive thrust of the distal extremity of the forelimb when the middle phalanges are flexed against the substratum. Secondarily, the long flexor muscles may also provide support to the volar aspect of the wrist during such vigorous locomotor activities as galloping and landing after a jump. Tuttle has suggested a number of other muscular mechanisms that are mechanically advantageous when seen within the context of knuckle-walking.

Related to such muscular contrasts is a series of bony features: for instance, there is the extension of the articular surfaces on to the dorsal aspects of the

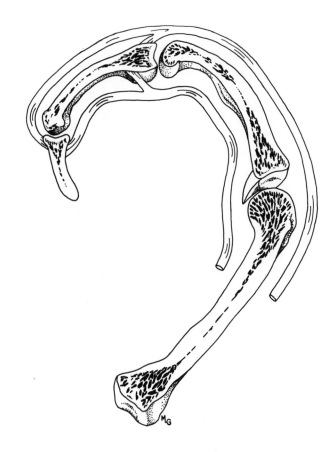

Figure 16. The hand as a flexible hook. A diagrammatic longitudinal section through the principal digital ray of the orangutan when placed in a typical hanging posture. Anatomical features are simplified and slightly exploded.

heads of the second to fifth metacarpals, thus permitting considerable hyperextension of the joints. Prominent ridges on the dorsal aspects of the metacarpal heads help to prevent displacement of these joints beyond the position of hyperextension. Oxnard (1969a) has shown that while the nature of the curvature of the phalanges of the orangutan is mechanically related to its use of the hand as a living hook, the special nature of the different curvatures, buttresses, and joint facets of the phalanges of the gorilla and chimpanzee are not inconsistent with mechanical efficiency in knuckle-walking postures and movement. As with the shoulder, so with the hand: careful and detailed functional and morphological studies, utilizing many specimens and checking functional possibilities, muscular anatomy, and osteological structure one against the other, can reveal many aspects of adaptation. Subsequent experimental stress analyses have allowed the examination of the structure of the hand to a further degree (see chapter 6).

The Pelvis in Primates

A series of studies has been carried out on the hip in the primates. Dissection of 141 specimens has shown many detailed correlations between soft tissue architecture and what is known about the function of the hip in locomotion. Such a

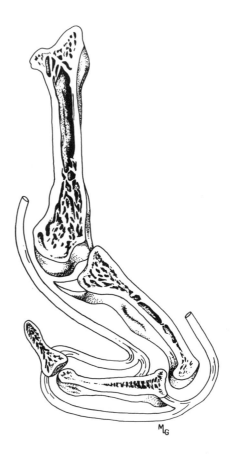

Figure 17. The hand in a knuckle-walking posture. A diagrammatic longitudinal section through the principal ray of the chimpanzee when placed in a typical knuckle-walking posture. Anatomical features are simplified and slightly exploded.

large number of dissections gives an idea of the variability both within and among different genera. As with the shoulder, a number of meristic features of the musculature of the pelvis are noted (e.g., the presence of a short head of biceps femoris in prehensile-tailed monkeys, saki monkeys, and the apes and man, but not in other forms). But the principal differences that seem to be of functional import concern the relative sizes of the different muscles and, in particular, of the different functional muscle blocks.[4]

Thus, in man, the four functional muscle units (extensors, flexors, adductors, and abductors) are of approximately similar size, forming as it were four guy ropes around a single column. In contrast, in most of the nonhuman primates the main bulk is centered around the extensors (especially) and the flexors, the

---

4. Two extensive studies of the pelvic musculature have appeared in the literature within recent years. One of these (Uhlman, 1969) is an excellent review of the morphology of the pelvic muscles throughout the primates, but it is without functional or evolutionary interpretation. The other (Stern, 1971a) provides much original information about the structure of the hip muscles in New World primates and discusses the findings in relation to the morphology of remaining primates and within functional and evolutionary contexts. The findings reported here are not at variance with those of these two authors.

adductors being considerably smaller and the abductors being extremely small (only one-tenth the size of the extensors). In the prosimians and in most anthropoids the bulk of the musculature is deployed for moving the limb in the craniocaudal direction with considerably less emphasis on side-to-side balancing movements; but it is also true that, in a few New World monkeys and in the apes, lateral movement is more important in climbing, and lateral musculature is correspondingly developed (Stern, 1971a). This is a brief summary of an extensive series of findings (Oxnard, 1966; Zuckerman, 1970) that will be reported elsewhere.

Information of this type is used to sort out from the many features of the innominate bone those which are likely on biomechanical grounds to be related to the muscular contrasts. Thus we may define osteometrically such features as (a) the relative caudal extent of the ischium, a measure possibly related to the length of the lever arm of the hamstring musculature (small in man but large in monkeys and apes); and (b) the placement of the iliac blade relative to the acetabulum, a measure of the position of origin of the lesser glutei, which are femoral abductors in man but which are stabilizers and rotators related to the power stroke (extension) in monkeys and apes. In addition to such measures of position of muscular attachments, it seems — again on mechanical grounds — that there are a number of other features which are related more to the relative disposition of the joints over which muscles act during locomotion. We may therefore also define such features as the dorso-ventral position of (a) the auricular facet and (b) the acetabular cavity, together with (c) the relative craniocaudal separation of these two joints. In all, nine such dimensions are defined and examined in 41 genera of primates, for most of which reasonable samples are available — a total of 441 specimens in all.

Univariate study of these dimensions suggests, albeit somewhat vaguely, many differences among the various primate groups (Oxnard, 1966; Zuckerman, Ashton, Oxnard, and Spence, 1967; Zuckerman, 1970). For instance, within the Anthropoidea there appear to be mechanically meaningful variations related to such functional differences as (a) the extent to which the hindlimb is subject to impulsive forces in leaping (e.g., many colobines in contrast to many cercopithecines); (b) the extent to which the hindlimb is capable of highly mobile three-dimensional movement (e.g., as in orangutans and howler monkeys); and (c) the extent to which vigorous protraction of the hindlimb is a necessary complement to propulsion during fast running (for instance, squirrel monkeys and patas monkeys). Within the prosimii also there are meaningful differences between (a) the mainly quadrupedal types (e.g., lemurs) which are, on the whole, similar to the quadrupedal Anthropoidea; (b) forms which hop (e.g., bushbabies and, in a different manner, sifakas); and (c) those which adopt unique slothlike patterns of movement (for instance, the angwantibo and potto).

However, the univariate analysis of these data shows clearly only one major feature: that is, the unique separation of the human pelvis from that of all others. Although many other interesting differences appear, as outlined above, the large number of genera available, the number of different dimensions to be examined, and the existence of some obvious and other less obvious correlations between one dimension and the next, all make it almost impossible to see what further information is enclosed in the data. Again, as for the scapular dimensions, some form of further analysis is necessary (see chapter 3).

We must be the first to point out that there are many other features of the primate innominate bone which have not been included in this study; some may also give information relating to locomotor features of the bone; others are in all probability related to quite different nonlocomotor aspects of the shape of the pelvis. Most features are of course liable to be related to any or all of the various functions of the pelvis. As Washburn (1964) has so rightly pointed out, "the form of the pelvis is a compromise between several sets of functional requirements, and the different sets may evolve with a considerable degree of independence." Nevertheless the attempt must be made to separate information from different features as far as possible. Only then can a synthesis of extant forms be attempted and the meaning of fossil structure more objectively elucidated.

Summary

As a preliminary step before attempting descriptions of the various mathematical, physical, and engineering analyses of different regions of the primate locomotor system, brief reviews of appropriate functional morphology have been presented. For more detailed information the reader must turn to original publications. But the simple descriptions of the anatomical features of the shoulder, hand, and hip just discussed are adequate for further elaboration. Each in turn leads to the more complex studies of later chapters.

An attempt has also been made to describe the general strategy of investigation so that parallels from one species to another within the Primates can be identified. This comparative approach is essential in evolutionary studies, and restriction to anatomical regions rather than study of whole animals provides background information that is necessary for the functional diagnosis of fossil fragments.

Relationships between function and structure that are anecdotal may be most important in initially channeling our ideas and in helping provide hypotheses worthy of testing. It must be emphasized, however, that such speculations are scarcely part of scientific methodology itself. Though subsequent study sometimes appears to corroborate such speculations, we must search especially diligently for their rejection. Functional-structural interactions may prove so complicated that several levels of hypothesis may be necessary for better understanding.

3 Multivariate Morphometric Analysis

Introduction

Once relationships between the mechanical aspects of behavior and the architectural features of muscles and bones have been outlined, it is possible to examine in considerable detail the functional morphology of individual bones and bony fragments. The fairly simple characterization of bone shape by means of a number of measurements can be refined by utilizing techniques that make allowance for differing sizes of specimens and that take into account correlation among characters within individuals and groups of individuals. Such multivariate statistical methods (factor, principal component, generalized distance, and canonical analyses, together with other related techniques) are capable of providing discriminations relating to many variables taken on many individuals within single or multiple groups. Some of the best examples of the use of these techniques in primate evolution are still confined to a number of the earliest studies (for example, Howells, 1951, 1957, 1966; Mukherjee, Rao and Trevor, 1955; Ashton, Healy and Lipton, 1957). These techniques are capable of providing (a) statistically significant discrimination between samples of biological materials, (b) clearer understanding of the relationships among the different measurements (original variables) taken on each object, and (c) diagnosis of unknown specimens.

Thus, even when there are only slight differences between samples in terms of individual original variables, multivariate analysis may show where the essential differences actually lie, something that cannot be achieved reliably by univariate methods. For example, the studies of Howells (1966) show that samples of data from sets of skulls can clearly define differences between various human groups. These analyses also indicate important dimensional variations, for instance that the greatest single separator in modern man lies in the cranial breadth, not, however, at its maximum on the vault but in the various measures of breadth of the cranial base. Furthermore, Howells is able to define within his many dimensions subgroups that may be linked in biologically meaningful ways; at the very least such exploratory use supplies hypotheses about the biology of skull form which can then be tested by other studies. Finally, such studies are capable of unequivocally placing certain unknowns (e.g., the Fish Hoek and Keilor skulls) within human groups, conclusions which are of distinct importance in the historic diagnoses of these forms (Howells, 1969).

Of additional interest is the use of these methods in those very cases where objects are obviously and significantly different and where some characterization of the degree and nature of the differences between them is required. In these cases it is highly likely that multivariate transformations will give opportunities for functional interpretation and possibly provide tests of hypotheses about functional adaptation. This use of the techniques has been carried out to a large extent in sciences other than biology (e.g., in sociology by Bock and Haggard, 1968) and, within biology, in areas other than the evolution of primates (e.g., in entomology by Blackith and Kevan, 1967). Among works dealing with the non-human primates the use of these techniques seems to be mainly confined to published studies on the shoulder (e.g., Ashton, Healy, Oxnard, and Spence, 1965;

Oxnard, 1967, 1968a, 1969b, 1970; Ashton, Flinn, Oxnard, and Spence, 1971) and to studies on the pelvis and hand currently in progress (Oxnard, 1972a).

The core of many of these techniques stems from the fact that whenever there are observationally difficult questions postulated by *correlated variables* in a "flat" (Euclidean) surface, space, or hyperspace,[1] they sometimes may be more clearly understood through answers resolved by the examination of *uncorrelated variates* in an appropriately "curved" region.

The precise techniques that may be adopted depend upon the particular questions that we wish to ask. In some cases our attention may be directed toward data taken either from what we believe to be a single group of specimens, or from a series of groups so entangled that they cannot be readily distinguished by eye. This is likely to be the case in examining data from subspecies, for instance, or geographic clines. At this level of hypothesis we may ask such questions as "what is the structure of the original variables within this single group of specimens?" Here the techniques include the general set of methods known as factor analysis (including principal component determination).

In an analogous way we may be interested in knowing the relationships that exist among several clearly defined groups (species, genera, populations, localities, time slices, etc.) where each group is believed to be relatively homogeneous and known to be different from the others. This pertains, for example, to a situation in which one is examining different extant genera of primates; for on the whole there is little argument among primate taxonomists about what constitutes the majority of the genera; most genera are established and "real." This may be much less so, however, at other taxonomic levels and with fossil materials; in some cases groups may be "real"; in others they may be figments of a typological train of thought; they may sometimes be due to deficiencies in the samples of specimens. Nevertheless, notwithstanding qualifications of this nature, it is statistically possible to analyze data when the question we wish to ask is "what are the relationships between known (or given) groups?" Such methods include those known as generalized distance and canonical analyses.

The complete understanding and use of techniques such as these require familiarity with matrix algebra (see Searle, 1966). The full technical description of such methods is a task for the biological statistician, and several adequate characterizations are now available (e.g., Seal, 1964; Blackith and Reyment, 1971). But it is of considerable value for many biologists to have a shorthand, geometric understanding of what the techniques can show and how they differ from one another. Such descriptions allow those who are not well versed in the techniques to be able to evaluate studies in which they have been utilized; in addition, these nontechnical synopses may attract those who do not currently understand the methods and may help them gain further insight and perhaps a desire to utilize them.

---

1. Regular space is high and wide. Hyperspace is just outside (Winsor, 1958).

The Structure of a Single Group

The simplest situation is one in which we have what we believe to be a single group of objects which can be defined by a number of variables. In other words, we may be able to describe the shape of each specimen in a group by a series of measurements. Any given specimen may be represented as a single point located in a multidimensional space. The coordinates of the point are the actual values of each of the measurements, and the many dimensions are the many different measurements themselves. In this system, a number of closely similar specimens (i.e., all of the same group) correspond to a number of points lying close together within the multidimensional space. Such a set of specimens may be thought of as a cloud of points. The problem then in investigating the structure of this single group is to understand the shape of the cloud.

A complication exists, however, in the fact that each original measurement does not necessarily give completely new information about a particular specimen; for instance, measurement of a second side of a cube tells us nothing about that cube that was not contained in the measurement of the first side. Accordingly each measurement can be treated as a new dimension in the multidimensional space only to the extent that it gives information about the specimen that is not already proffered by the other measurements. It is the question, "how much of each measurement is presenting new information about the specimen?" that can be answered by the manipulations of multivariate statistics. The answers often show that an apparently complex, multidimensional space may be reduced to a simpler, few-dimensional domain with the loss of very little information and a great gain in understanding.

A simple example of how the technique works may be seen by considering two-dimensional (and therefore easily visualized) examples. Thus figure 18 shows a single cloud (population) A with the positions of each of its component specimens (a's) plotted in terms of two measurements (x and y) taken on each specimen. If we examine the population in terms of only one measurement, say x, then the specimens are relatively close to one another (projection of a's on the x axis). This is similarly the case if we examine the population in terms of y. However, if we look at both dimensions together, we see that the specimens (the a's) are actually spread out in an elliptical distribution. If the two variables are not correlated at all, then the structure of this single group of organisms is an ellipse whose major and minor axes are parallel to the measurement axes (i.e., x and y). The difference in the major and minor axes is related to the variation of each measurement.[2]

When, however, the two variables are correlated to some extent, then the ellipse is found to be rotated in relation to the x and y axes. The nature of the rotation is related to the extent of correlation, while the shape of the ellipse is related to

---

2. If we scale the x and y axes so that they become transformed in terms of the variance of the original measurements, then we obtain a circle, a concept that we will utilize later (figure 20).

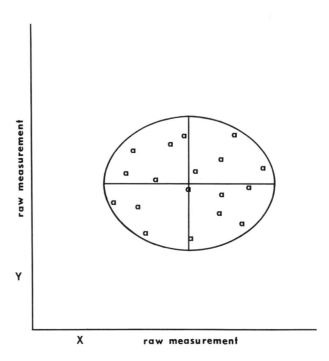

Figure 18. Bivariate plot of two uncorrelated variables (x and y) taken upon the specimens (a's) of a single population (A).

transformed variations of the x and y measurements (figure 19). The shape of this new ellipse can also be characterized by its major and minor axes (i.e., in terms of the projection of the a's on the two new axes, λ' and λ'', lying at some angle to the original axes). It is of interest that the first of these axes, λ', gives the maximum spread (major axis) of the group, while the second gives a considerably smaller spread (minor axis) which is independent of the first. If we extrapolate to the case where many original variables $x_1, x_2, \ldots x_n$, have been taken on each specimen, then it is likely that the number of new independent axes expressing statistically significant information will be considerably smaller than the number (n) of original measurements made on each specimen. These axes are referred to as components.

The equivalent algebraic manipulation is carried out so that the first component gives the major part of the structure of the cloud; each of the later components takes up, in turn, the greatest possible fraction of the new residual variation. These components are represented by the principal axes of ellipsoidal (or hyperellipsoidal) figures in the geometrical example. The procedure yields a number of components equal to the number of observed variables; but as the major part of the information is contained in earlier components, it may sometimes be possible to drop later components as having little content. It is in this way that the problem can be simplified.

Moreover, because of the nature of the statistical manipulation that is carried out when this is done, the new components into which the original correlated variables are transformed are statistically independent. Each component con-

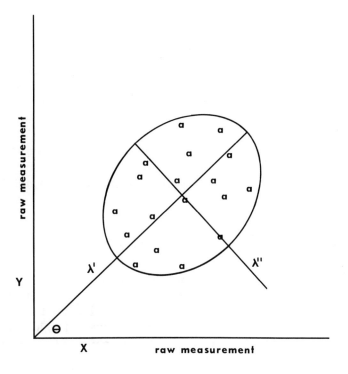

Figure 19. Bivariate plot of two correlated variables (x and y) taken upon the specimens (a's) of a single population (A). The new λ' and λ'' represent (in a geometric simulation) the principal components of A.

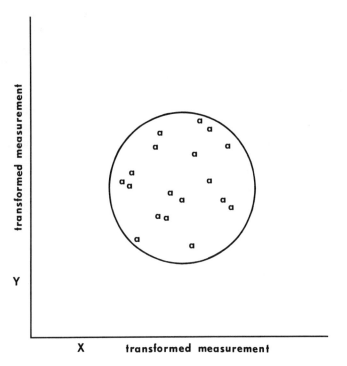

Figure 20. Equalization of variances. A diagrammatic representation of figure 19 when the axes have been rescaled so that x and y represent variables transformed to equalize their variation.

tains its own item of information about the shape of the cloud of points. This then is a geometric representation of the more complicated procedure of principal component analysis.

In other words, in an apparently homogeneous group, the several characters measured in an investigation may be factorized to reveal the underlying components of their variation. This process of factorization is intellectually (although not algebraically) similar to the familiar decomposition of a quadratic expression into its roots.

In algebraic terms, the variation and covariation within the single multidimensional cloud (represented in the simple example by the scatter of the a's within the ellipse A) is expressed by the matrix **W**. When this matrix is arranged in the determinantal equation

$$|\mathbf{W} - \lambda \mathbf{I}| = 0$$

then the solution is a set of roots, $\lambda', \lambda'', \ldots$. By this process the first component, λ', is the algebraic equivalent of the rather simpler geometric axis λ' in the example. Here the matrix **I** is a unit matrix of the same rank as **W** but with unity in the leading diagonal and zero elsewhere. This matrix therefore represents the variation of each measurement normalized to unity and zero correlation of any measurement with any other measurement, and it is in this way that the maximum uncorrelated variation is obtained in each principal component. If the data are scaled in some other way, the solution is different. In other words, the principal component analysis is not scale free.

The primary objective in principal component analysis is to account for as much of the total variance as possible; because of the nature of the above equation, most of the information in which we are interested will be in the earlier principal components. When the original variables are actual measurements, and when most of the loading factors are positive, then the first component will, in general, be an indicator of overall size. The next few components may carry considerable information about the data that are not related to overall size; this separation may be useful. The lattermost components may contain very little information and may represent "noise" within the data. Hence a useful feature of this technique is the "incomplete solution" in which some components are disregarded or discarded.

One can also relax the condition that these components are at right angles (independent of one another), and they then become the factors of factor analysis.[3]

---

3. In addition, there are a number of other modifications that can be applied; axes may be rotated to other positions for particular reasons, or axes may be deliberately generated that are not orthogonal (i.e., oblique) and therefore not truly independent. Such rotated or oblique solutions seem to have been especially used within the social sciences. In biology they have been used relatively little although recently a number of studies have been appearing in which their value is demonstrated (e.g., Gould, 1969). They may well have an important place in the examination of biological data. However, there is some question as to how we derive the information determining why particular rotations are used; circular reasoning must be avoided when applying particular rotations and obliquities to biological data for analytical purposes.

The Interrelationships of Many Groups

When the question that we wish to ask relates to the investigation of a number of different groups of specimens, then the problem is somewhat more complex. In this case it may be the equivalent of the geometric separation of the populations A, B, and C, as in figure 21. Here each population is defined by the positions of each specimen (a's, b's and c's). In this case, however, we are not interested in the structures within the groups, and thus the measurements, x and y can be scaled. One such technique, for instance, is to change the raw measurements into standard deviation units. If the data are reasonably normally distributed, the effect of this is to transform the ellipse for each single group (the elliptical structure of figure 18) into a circle (figure 20).

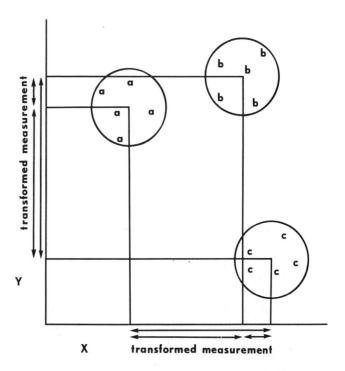

Figure 21. Bivariate plot of three populations. A diagrammatic representation of the positions of three populations, A, B, and C, when viewed as the bivariate plot of measurements x and y (transformed as in figure 20 to equalize variations) and taken from the specimens (a's, b's and c's) in each population. The positions of the populations in relation to the transformed measurements are shown.

As in the previous example, projections can then be made upon the x and y axes. But in this case what is projected is the center of each cloud of specimens rather than the position of each specimen itself. This gives the distances of the three clouds from one another when viewed in terms of either measurement x or y respectively. These separations are not necessarily the best that can be achieved. Because we can view both the x and y dimensions together, we know that the real separation of the groups is somewhat greater.

For instance, one measure of the separations of the three groups is given by the three distances of each from the others (figure 22). These distances consist

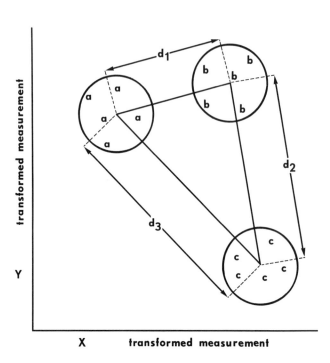

Figure 22. A diagrammatic representation of the process of generalized distance analysis performed upon the data of figure 21; d_1, d_2, and d_3 represent the appropriate distances.

of the full separations between the three clouds, but they exist in two-dimensional space (three points define a plane). We shall return to this concept.

A second way of looking at the clouds is to perform the maneuver of the principal component example and reproject the centers of the clouds onto that axis which most separates the three groups (figure 23). This is the new axis λ' (known, when derived by appropriate algebraic techniques, as a canonical axis or variate). However if we do this, we should also scan the projection of the clouds onto that axis which is at right angles to λ', that is, λ''. This also separates the groups, but to a lesser extent than the first. (If, of course, the groups happen to lie in a straight line, the second axis will provide no separation at all, as the first of the new axes would be the given straight line.) Similarly, if we extrapolate to the complex case where a large number of measurements are taken on each group, it is likely that the number of new variates that give statistically significant information will be considerably fewer than the number of original measurements. And again, whereas the original measurements are all correlated with one another, the new canonical variates are, by definition, independent.

It is possible to add more and more groups, but as one does so the procedure becomes more and more complicated. However, assuming for the moment that the number of measurements taken on each specimen is high, the means of two clouds always lie on a straight line (i.e., can be represented by positions along a single canonical variate or axis). The means of three clouds always lie in a plane

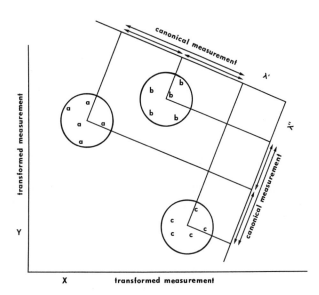

Figure 23. A diagrammatic representation of the process of canonical analysis when applied to the data of figure 21. The new axes λ' and λ" represent the appropriate canonical axes. The positions of the populations *A*, *B*, and *C* in relation to the canonical axes are shown.

(i.e., only two newly constructed variates are required to visualize their relative positions). For four groups, the means lie within three-dimensional space, and so on. This relates to the well-known fact that comparisons between two means represent only one degree of freedom; among three means, two degrees of freedom; among four means, three degrees of freedom; and so on. It tells us that when we are discussing the relationships among a number of groups, the dimensionality of the relations is important. When there are more variables (measurements) than groups, the number of new canonical variates will be limited to one less than the number of groups.

A second limitation to this situation also holds. Thus, if instead of a multivariate problem only one variable is available, then all the differences among many groups will be spread along a single axis representing that variable. Similarly, with only two variables all the differences must lie in a plane. So in general, the effective number of new discriminating canonical variates cannot be more than the minimum number of variables; and we have already shown that the number cannot be more than one less than the number of groups. In practice, however, the number of groups and variables will both be large, and hopefully the number of canonical variates that adequately summarize the data will be considerably less than each.

This then is the geometric representation of the more complicated process of the canonical analysis of a number of clouds in a multidimensional space. In algebraic terms the variation and covariation *between* and *within* the different multidimensional clouds (represented in the simple example by the scatter of *a*'s, *b*'s and *c*'s *between* and *within* the populations *A*, *B* and *C*) is expressed respectively as the matrices **B** and **W**.

The solution of the determinantal equation

$$|\mathbf{B} - \lambda \mathbf{W}| = 0$$

then yields as many roots (λ', λ'', . . .) as there are characters, though not all the roots are necessarily of interest. As in our simple example, the number of roots may be smaller if the number of populations is less than the number of characters; in this case there will be one less root than the number of groups. The first root gives that linear combination of the various measurements which best discriminates between the populations under consideration, and it is tested for statistical significance. The second provides that linear combination of the measurements which is uncorrelated with the first and which best discriminates the populations using the remaining differences between them. Further linear combinations are derived in the same way, each being uncorrelated with those already computed. The linear combinations so formed are called the canonical axes or variates.

Using the mean value for each measurement on each population under examination, we can compute the value along each canonical axis for the mean of each population. The values so computed can be plotted linearly (for one canonical variate), or, more generally, in k dimensional space if k canonical variates are computed. This gives a convenient method for assessing the relative "positions" of the various populations.

The canonical variates may be examined with significance in mind. First those that are statistically significant are obviously worthy of consideration. But of that number some may give information apparently of such small degree that they may be judged to be of less biological importance than earlier variates. On the other hand, early variates may contain large but (within the context of particular biological problems) uninteresting information, such as differences in overall size, or marked phenotypic plasticity. Later variates may contain smaller portions of the data but may be more interesting in terms of functional, genetic, or taxonomic significance (previously masked by larger phenotypic or size variability). However, judgments of this nature are not mathematical and should be examined with care.

Each variate is independent of (orthogonal to) the other canonical variates within the multidimensional space defined by the matrices. Thus the contrasts in form that they represent are mutually independent. It may be useful to point out that the orthogonality of the canonical variates is in a multidimensional curved (Riemannian) space and not in the flat (Euclidean) space in which they are normally drawn for viewing. Thus in Euclidean space they may make angles of other than 90° with one another. However, as the diagrams on plane paper (Euclidean) are in effect projections from the Riemannian space, an orthogonal representation seems appropriate. It follows also that there is not necessarily any direct comparison of principal components with canonical variates. Comparability is scarcely to be expected since the principal components are mutually orthogonal in multidimensional Euclidean space, while canonical variates are orthogonal in the Riemannian space.[4]

If such examinations suggest that the derived canonical axes seem to be conveying information of biological import, then a number of tests *must* be made. For it is clearly somewhat speculative to suggest mathematico-biological interrelationships of this nature. Most of the tests which should be applied are based upon the use of (a) other analytical techniques, (b) examination of new data from the same specimens, (c) examination of data from new specimens, (d) independent experimental tests of the nature of the biological association suggested, and so on. These are all discussed in other sections of this book. But one immediate test that follows rests in the discovery of the nature of the information that is contained within particular canonical variates. This can be achieved by the calculation of the fractional contributions of each original measurement to each canonical variate. There are a number of different ways in which this can be done; clearly, however, it is an important step in understanding the degree to which a given canonical axis may represent an aspect of the difference between the groups of specimens.

It has been pointed out above that rather than representing the positions of the clouds on the canonical axes, these positions may be represented in terms of their actual distances from one another within the full multidimensional space. This is generalized distance analysis (figure 22). The results of this technique are easily understood when the number of canonical variates does not exceed three; in that case, the distances can be absolutely and correctly represented by a two- or three-dimensional model. The results can still be understood if the number of canonical variates exceeds three, as long as the number of groups is small; for although the true picture cannot be physically modeled, the numerical values can easily be examined by eye. And finally, even when there are many groups the method may still be useful when the point of interest is the position of all other groups in relation to one special group (for example, in studies of the evolution of man, a special group might be man himself or possibly some rather relevant fossil). In this case, the data can be represented by a connected graph, though care must be taken to recognize that distances of nonconnected items in such a graph are *not* equivalent to "true" generalized distances.

However, when the affinity of a large number of groups is represented by generalized distances, visualizing them becomes most cumbersome. For instance, in

---

4. Nevertheless, Howells (1972) in a fascinating study applying the two forms of multivariate statistics (factor and discriminant analyses) to the same body of data, has obtained similar results indicating that interpopulation differences in the skulls of recent man involve the same morphological patterns as does individual variation within populations. The morphological patterns do not confirm classical distinctions such as cranial and nasal indices. Rather, they consist of general differences in a large number of facial and cranial base features. Although Howells points out that there are limitations in the methods relating to the essential indeterminacy of the transformed variates, he also indicates the great flexibility and breadth of possible uses of transformed variates that makes them an increasingly useful set of tools in examining data and testing hypotheses about variation, form, and function.

the studies on the primate shoulder, the examination of such distances involves the examination of a 41 by 41 matrix of generalized distances. It is here that the combined use of generalized distance and canonical analysis may be useful. For the generalized distances may be plotted against the background of directions supplied by a small number of canonical variates. This may demonstrate just how little or how much information for each group resides in later canonical variates. This technique may be especially useful when allied with cluster analysis based upon generalized distances, and when data from unknown forms are being interpolated into analyses for the purpose of characterization (see chapters 5 and 7).

We can summarize this discussion of multigroup analysis by saying that generalized distance methods attempt to supply the absolute distances of various groups from one another within a curved multidimensional space. Canonical analysis attempts to superimpose upon this multidimensional space a set of orthogonal axes that have the property of producing as much of the separations as possible in a single (first) canonical axis, then the majority of what remains in a second axis, the major part of the residual in a third, and so on; the object is, hopefully, to get enough of the separation in a sufficiently small number of axes to be able to examine the relationship in a bivariate plot, a three-dimensional model, or a small number of these combined. In this way the task is made easier for the human eye. Thus generalized distance methods "change" large numbers of original correlated variables into single distances between groups. Canonical analysis transforms large numbers of original correlated variables into small numbers of new uncorrelated canonical variates which provide succinct synopses.

Biological Meaning of Mathematical Parameters

These various techniques (and a host of others related to them) are available for the examination of biological data. Though it is always possible to find mathematical relationships between groups—and this was certainly the initial use of the methods—it may also be possible to impute nonmathematical meaning to the structure that is revealed. This has been done by a number of workers generally outside the realm of primate anatomy and evolution (e.g., see especially Blackith and Reyment, 1971). In entomology Blackith and coworkers show that biological meaning appears to hold in a number of different investigations. For instance, in studies of grasshoppers of the genus *Chrotogonus* (Blackith and Kevan, 1967), a first canonical variate seems to separate groups in relationship to sexual dimorphism and a second with regard to alary polymorphism. In the mirid bug *Plagognathus,* Blackith (1965) suggests that a first axis separates the forms in relation to sexual dimorphism, a second in relation to influence of host upon growth, and a third in relation to geographic variation. For morabine grasshoppers, Blackith and Blackith (1969) find that sexual dimorphism, rather than being oriented with a first canonical axis as in many other forms, is oriented obliquely so that it is expressed in terms of both first and second canonical axes. Axis one displays size variation, and axis two seems to reflect the degree of attenuation of form. In this particular study approximately four canonical variates are obviously relevant, and information of some degree is still being obtained from canonical axes as high as the seventh.

Outside the realm of biology, Blackith (1963) has also shown meaning (in this case, literary) apparently to exist within mathematically determined principal components. In his study of Latin elegaic verse a first principal component relates to the fluctuating degree of organization of words. Thus on this axis large negative values correspond to relatively disorganized poetry and characterize the work of poets before (such as Catullus) and after (such as Martial, Avianus and Rutilius Namatianus) the Golden Age. In contrast, such Golden Age poets as Propertius and Tibullus demonstrate relatively tightly organized poetry with large positive values in this axis. In this study the second principal component seems to be related to the differential control of elision frequency; this is an important element of literary style peculiar to each writer and has been used to challenge authorship. A third principal component displays a considerably smaller amount of the total variability but in general proves of great value as a discriminant of Latin elegaic verse. Mean syllable number and elision frequency varying together characterize this third principal component, which can be used to predict the date of a poem in a way that is apparently refractory to counterfeiting. The score in this component rises to a maximum during Augustan times and then drops off steadily (although while the stylistic features of Latin elegaics were evolving during the first centuries of this era, the drop-off is naturally sharper than during the long static period from the decline of Rome to the present day).

Within the field of anthropology and primate anatomy some attempts have been made to impute biological meaning to mathematical parameters. We must emphasize, however, that, especially in anthropology, the nature of the problems (investigation of subspecies, races or geographic variants that are closely similar; the investigation of regions like the skull which consist of anatomical complexes with a large number of functions closely interwoven) is such that we do not expect unequivocal biological meaning (Howells, 1969, 1972). In these fields the techniques are more often used as searching tools rather than as hypothesis-testing mechanisms. Nevertheless, this caveat notwithstanding, a number of studies have pointed to factors of size and body linearity versus laterality and robustness. They have shown that limb growth is apt to be differentiated from trunk growth; that intralimb proportions may be expressed in separate factors; that the head constitutes a system or systems apart from the rest of the body. Within the head a number of distinctions have arisen: particularly of general size, of relative brain size, of facial length, and so on (Howells, 1968, 1969, 1972). Hiernaux (1963), utilizing generalized distance statistics for analyzing data from tribes in the eastern Congo and Rwanda Burundi, found gradients running east and west. These appeared to be open to interpretation either as environmental in cause (shade forest versus upland) or as due to genetic intermixing in different degrees of the various parent stocks. A number of studies have thus shown biological factors of this sort to be expressed in terms of mathematical parameters.

Ideally such factors may be found to correspond to distinguishable (if coordinated) events, for instance, in development to hormonal, genetic, or field control, or in adaptation to environmental variables. If, in fact, this turns out to be the case, then scoring a component or a variate for an individual or group provides a method of assessment which should be a powerful weapon in the investigation of shape difference and of how it may have arisen. The biological reality and the exact specifications of such axes or factors is not yet clear, though it is difficult to doubt their existence. Too little has been done to test correspondence of factors from different studies.

This is one of the most exciting possibilities in these approaches: the opportunity to begin to identify factors, components, variates, which will appear if the correct questions are being asked, regardless of the method used. Such robust parameters, obtained independently of the method, are the ones that should be considered in making substantive conclusions about the nature of the data domain. All others should remain tentative and be subject to further study. Parameters which are robust in this manner are surely not artifacts of the particular methods used and should not be subjected to the same doubt as those obtained from a single method of analysis. It is true, however, that as parameters appear which have a good deal of similarity over methods, but which are nonetheless not identical in all respects, some elasticity will enter the picture. This cannot be predicted and will not be known until such time as the comparison of parameters over methods becomes a common practice.

These various features of the methods can be demonstrated to a considerable extent in studies of data from the primate shoulder.

Investigation of the Primate Shoulder: "Locomotor" Dimensions

Investigations of the primate shoulder are virtually complete. Of greater interest than the separation of different primate genera is the *nature* of the separations which these investigations have shown. The study resolves itself into the separation of forty-one genera of primates one from another, when each genus is defined by nine dimensions taken on the shoulder girdle, a total of five-hundred and fifty-one shoulder girdles in all. The particular osteological dimensions which were chosen appear to be related to muscular anatomy viewed in the light of presumed function of the shoulder in locomotion in the different primates (as has been described in chapter 2). Canonical analysis of this considerable set of data suggests that the primary differences among all the nonhuman primates may be defined by only two canonical variates or axes; these new variates seem to arrange the different forms in ways that make sense when seen in the light of different locomotor patterns (Ashton, Healy, Oxnard, and Spence, 1965; Oxnard, 1967).

Thus the first canonical variate apparently arranges the nonhuman primates in relation to the extent to which the arm is raised and bears, or seems capable of bearing, tensile forces. The second variate seemingly sorts out the animals according to the extent to which they climb and live in trees. Man is uniquely separated from the nonhuman primates by a third canonical variate which may be thought of as correlating with the unique functions of the human shoulder girdle.

In other words, it seems as though biological meaning may be imputed to the mathematical derivatives; and though this is normally a somewhat speculative type of suggestion, in this particular case the likelihood that the speculation is correct has been greatly increased by a number of tests.

Investigation of "Residual" Dimensions

Another series of eight measurements is available from the same set of animals. These "residual" measurements differ from the above in that they are chosen (a) so as to correlate mathematically as little as possible with the first nine, and (b) so as to have as little mechanical relationship as possible to the earlier group (as far as can be judged from usual anatomical inference). This results in these eight measures being a conglomerate of dimensions; some are similar to standard anthropological indices; some are chosen because they provide obvious measures of scapular elements without particular relevance to functional meaning; some are chosen in an almost haphazard manner. They are described in detail in Ashton, Oxnard, and Spence (1965).

Canonical analysis of these data suggest that similar information is inherent in the anatomical shapes, but to a somewhat less clear extent. This is perhaps to be expected, for although it is undoubtedly possible to choose certain dimensions as reflecting function in a circumscribed region to a greater degree, most other dimensions should also be correlated to some extent with the former and therefore also with function.

Thus the bivariate plot of the first and second canonical axes of this new study provides, first, a generally linear arrangement of all primates that parallels the arrangement found in the canonical analysis of the nine locomotor dimensions. The most quadrupedal forms (in which the lowered shoulder bears primarily compressive forces) appear at the right-hand end of canonical axis one, and the most acrobatic forms (in which the raised shoulder bears tensile forces to a greater degree) lie at the opposite end of this axis. This separation, as can be seen from figure 24, is not as obvious as that shown by the analysis of the locomotor data set (figure 28).

The information given by the second canonical axis of the locomotor analysis (separating forms according to the degree to which they are arboreal or terrestrial) is not obvious here. But appearing in the second axis of the residual analysis is information relating to the separation of certain prosimian forms from one another. Thus the relatively quadrupedal prosimians, in which the shoulder bears weight principally by compression, are separated by this axis from the relatively acrobatic prosimians, in which the shoulder sometimes bears tensile forces. This is of considerable interest because these two separations are achieved together by a single axis in the previous analysis of the locomotor dimensions whereas in this new analysis two independent axes are required; yet it is clear that information about these same biological factors is inherent in the new data. This is confirmed when we examine, in the residual analysis, the separation of man from other primates; this is uniquely achieved as in the previous locomotor analysis by a third canonical axis.

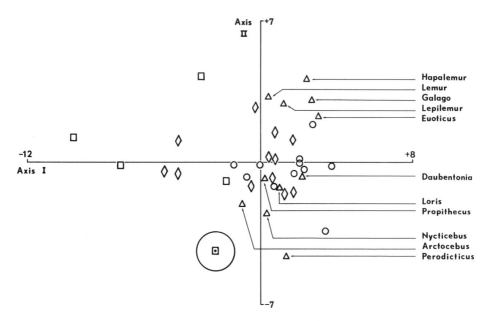

Figure 24. Bivariate plot of the first two canonical axes in the analysis of the eight residual dimensions of the primate shoulder girdle. Squares represent apes and man; diamonds represent New World monkeys; circles represent Old World monkeys; triangles represent prosimians. The position of man is marked by the circle around the mean. This circle also serves as a scale marker, having a radius of one standard deviation unit. The most marked separation that is achieved appears to be that between the named groups of prosimians. (The position of *Daubentonia* is spurious; see chapter 7.)

Canonical Analysis of "Combined" Shoulder Dimensions

The combination of the two sets of measurements has now been carried out (Ashton, Flinn, Oxnard, and Spence, 1971) and the canonical analysis of the total of seventeen dimensions of the shoulder is now available. Because this new analysis gives somewhat more information than either subset analyzed alone, the full information is figured here.

The results show some, but not extensive, differences from the picture that was obtained from the analysis of the nine locomotor dimensions alone. Certainly the addition of the new dimensions has not negated the previous conclusions (see figures 25, 26, and 27). In fact, the reverse is very much the case: it seems remarkable just how few dimensions appear to be able to give an adequate picture of the situation.

Thus examination of the first canonical axis (as seen from the horizontal axis of figure 25) demonstrates very clearly the relationship of the placement of forms in this axis with information relating to the extent to which the forelimb is capable of being raised above the head and of bearing tensile forces. There is a clear spectrum running across the page from patas monkeys through squirrel monkeys,

Multivariate Morphometric Analysis

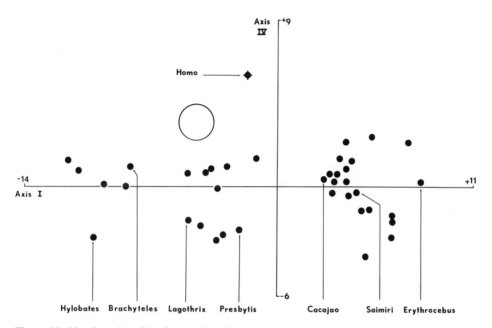

Figure 25. Bivariate plot of the first and fourth canonical axes of the analysis of the combined locomotor and residual dimensions (seventeen variables in all) of the primate shoulder girdle. The circle has a radius of one standard deviation unit. Selected genera are named. The figure illustrates the spread of the genera along axis one in a way that seems to correspond to the degree to which the arm may be raised and is capable of bearing tensile forces. It also demonstrates the unique separation of man in axis four.

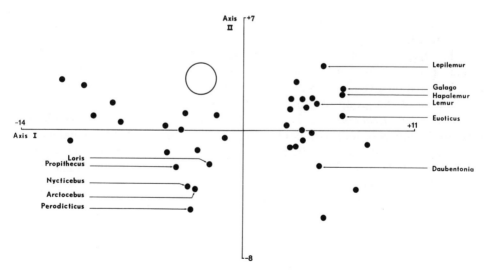

Figure 26. Bivariate plot of the first and second canonical axes of the combined (seventeen-dimensional) study of the primate shoulder girdle. The circle has a radius of one standard deviation unit. Selected genera are named. The figure illustrates the separation of the Prosimii that is achieved by axis two.

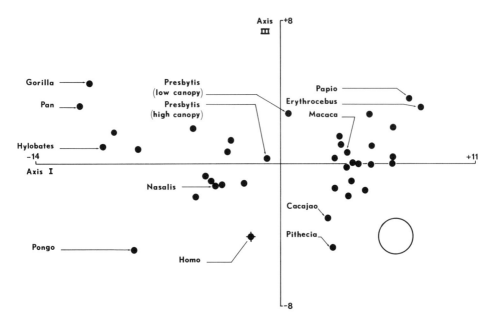

Figure 27. Bivariate plot of the first and third canonical axes of the combined (seventeen-dimensional) study of the primate shoulder girdle. The circle has a radius of one standard deviation unit. Selected genera are named. The figure illustrates the separation of the Anthropoidea seemingly according to the extent to which individual genera are terrestrial or arboreal; this is achieved by canonical axis three.

uakaris, langurs, woolly and woolly-spider monkeys in sequence, finally terminating with gibbons. Genera which have not been named lie in intermediate positions that relate very well to this concept (for specific details see Ashton, Flinn, Oxnard and Spence, 1971).

The separation that is achieved in the second canonical axis of the 17-dimensional study is given in figure 26. It is fascinating in that it includes in large degree the information given by axis two of the residual study alone—that is, the further separation of the group of prosimians, various lemurs, and bushbabies (*Hapalemur, Lemur, Lepilemur, Galago* and *Euoticus*), the shoulders of which bear mainly compressive forces, as contrasted with such forms as lorisines and indriids (*Loris, Nycticebus, Arctocebus, Perodicticus,* and *Propithecus;* see figure 24), in which the mode of locomotion is such that the shoulder sometimes also bears tensile forces, as in hanging or clinging. This information is also included in the locomotor analysis but to a much less marked extent.

The examination of the third canonical axis in the study of all seventeen dimensions reveals information similar to that in the second canonical axis of the locomotor study alone; that is, the separations achieved in this axis relate to the extent to which the animals are arboreal or terrestrial. Thus we can see from figure 27 that if we restrict our examination to the forms at the right-hand end of axis one, then they are generally arranged with the most terrestrial species, such as baboons and patas monkeys, in the most positive positions; semiterrestrial forms like macaques, intermediate, and highly arboreal forms, such as uakari and saki monkeys, in the most negative locales. In the intermediate region of axis one, a similar though less marked separation from positive to negative is found from the relatively least arboreal low-canopy langurs (e.g., *Presbytis entellus*), through high-canopy langurs (e.g., *P. kasi*), to highly arboreal forms like the proboscis monkeys. At the left-hand end of axis one this sequence is repeated from the highly terrestrial gorilla and somewhat less terrestrial chimpanzee through to that almost exclusively arboreal acrobat, the orangutan.

Finally the information that is given by this study in terms of canonical axis four relates to the unique position of man. This is shown in figure 25, and is apparently similar to canonical axis three of the locomotor study. Detailed examination of the residual study also shows, in a less clear way, that man is different from all nonhuman primates.

The remarkable correspondence between these three studies, not only in terms of the contained information (that we would expect from looking at data taken from the same specimens) but also in terms of the correspondence of information in specific canonical axes, suggests that the biological data are genuinely arranged in this relatively clear-cut manner so that they force (as it were) the mathematical choice of the canonical variates in these specific ways. The full degree of the correspondence is a little difficult to see from the bivariate plots just presented for full study. It can be seen most clearly, however, from the following comparisons of individual axes across the different studies.

Thus figure 28 shows the remarkable concordance between the information supplied by the first canonical axis in all three studies; the relationship to the biological speculation about compressive or tensile forces in the shoulder, and about a lowered or raised arm, is most marked.

Figure 29 shows the excellent equivalence between the second canonical axis in the locomotor study and the third of the combined study relating to degree of terrestriality or arboreality of the different forms regardless of positions in the first axis. The correspondence is not complete—note the somewhat different positions for species such as *Pan* and high-canopy langurs. This latter information seems to be foreshadowed in the residual data when seen in its entirety (e.g., as shown by cluster analysis of generalized distances), but is not apparent in any single canonical axis.

In like manner, the information that is clearly present relating to the separation of the prosimian groups in the second axis of the residual study is also clearly evident in the second axis of the combined investigation. This information seems also to be apparent, although to a much less degree in the second axis of the locomotor study (figure 30), and in the totality of the combined study this separation is also evident in yet another axis (three), as can be seen from the bivariate plot also shown in figure 30.

This somewhat confused arrangement of prosimian forms in relation to the canonical axes may well relate to the fact that the major number of genera examined are anthropoids and these forms may thus overload the analysis so that the axes are more pertinent to separations among anthropoids. That this appears to

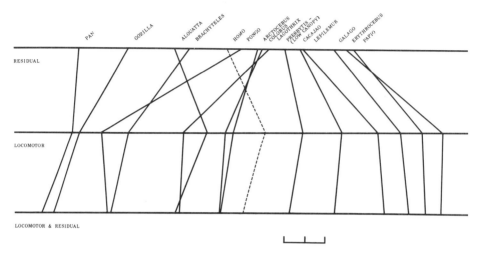

Figure 28. A comparison of the relative positions of selected genera in the first canonical axes of the three separate analyses of the shoulder data: residual (eight-dimensional), locomotor (nine-dimensional), and locomotor plus residual (seventeen-dimensional) studies. The general conformity of the rank orders of the genera indicates that similar information is being conveyed by each study. At the same time, the rank orders are not identical, showing differences relating to the different suites of measurements analyzed.

be so can be seen from additional canonical analyses performed on the data from the Prosimii alone, where a single axis suffices to produce maximal separation of the groups of genera. Presumably when prosimians are examined together with anthropoids, the single axis separation within the Prosimii alone must lie at some angle to (be nonorthogonal to) the major anthropoid separations. This may well be revealed by utilizing studies where rotated or oblique (nonorthogonal) axes are sought among the data (e.g., Hall, 1967); such studies have not yet been performed on these data.

Finally, the information about the unique separation of man from all other primates, given especially by the third axis of the locomotor study and mirrored to an even greater extent in the fourth axis of the combined study, is displayed in figure 31. This shows us only too clearly, that although these are separate canonical analyses, the same information is being presented by these two axes. Not only is the separation of man appropriately obvious, but also, even among those forms that are not well separated, rank orders along the axes are extraordinarily similar; this must represent almost identical views of the total multidimensional figures.

Study of the Shoulder in Nonprimate Mammals

A third test of the reality of the biological meaning apparently associated with the canonical axes involves the examination of the original set of dimensions, using the original canonical loading factors but applying them to a totally different series of animals, namely various nonprimate mammals. Here again, canonical analysis appears to confirm the biological meanings of the parameters in a manner

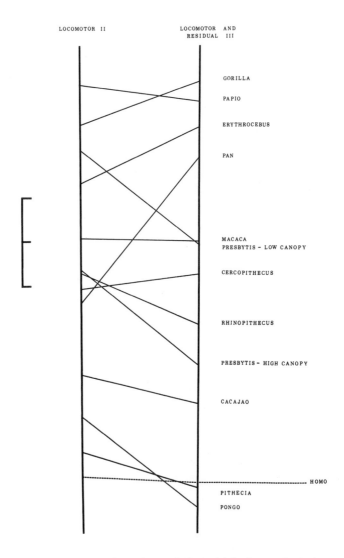

Figure 29. A comparison of the information conveyed by the second axis of the locomotor (nine-dimensional) and third of the combined (seventeen-dimensional) studies. In both, the forms are separated according to the extent to which they seem to be terrestrial or arboreal. The rank orders of the selected named genera are similar across the two studies, though in some cases (e.g., *Pan*) there are considerable divergences between positions of particular genera.

similar to the primates for all those mammals (arboreal) for which the analysis is appropriate. Thus nearly all the arboreal mammals lie in the same part of the canonical space as do the primates. Furthermore the separation of different forms by canonical axes one and two seems to hold similar meaning. Animals which hang (and in which therefore the shoulder is subject to tensile forces, e.g., arboreal sloths as compared with arboreal dwarf anteaters) are separated by the first canonical axis (figure 32). The most highly arboreal animals are defined by the second axis (e.g., the more arboreal dwarf anteaters as compared with the terrestrial giant anteaters; also, flying squirrels and tree squirrels as compared with ground squirrels). Close examination of the results suggests similar parallels in many mammalian groups (figure 33).

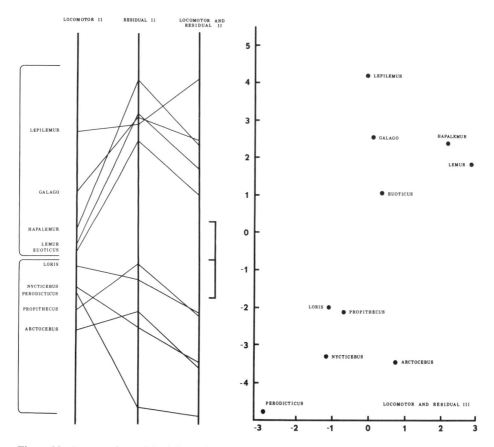

Figure 30. A comparison of the information conveyed by the second axis of all three studies with reference to the positions of the prosimian genera. Again the information conveyed is similar in each case, though it seems clear that it is demonstrated considerably better by the residual study than by the locomotor study. This confirms the suggestion (Oxnard, 1967) that information about prosimian separations seemed to be contained in the residual study. The third axis of the combined study also provides related information as is seen from the bivariate plot on the right-hand side of this figure. This suggests that the prosimian separation may actually lie at some angle to the axes that are produced in this study; these latter are presumably forced upon the prosimian data by the much larger number of anthropoid forms in the analysis. A subsidiary canonical analysis of the prosimian data confirms this because it produces this separation in a single axis.

As with the primates, these two separations tend to go somewhat together so that the average pattern of difference is a generally oblique linear arrangement; again, however, there is a sufficient number of groups which fall away from such a linear arrangement to demonstrate that the apparent biological dependence is not frozen or tied—it has been relaxed in some cases.

Because man is the only primate separated by canonical axis three, we are now especially interested in any separations that may be achieved by this axis among

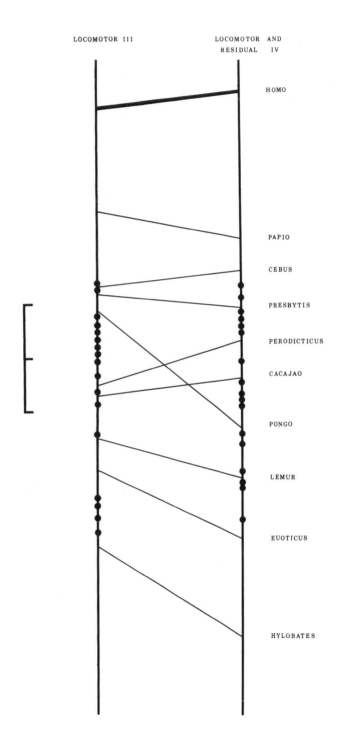

Figure 31. A comparison of the information conveyed by the third axis of the locomotor (nine-dimensional) study and the fourth axis of the combined (seventeen-dimensional) study. *Papio* and *Hylobates* are the extreme genera of the nonhuman primates in these axes. The figure demonstrates that man is truly very separate and that all the other forms are closely crowded in approximately similar rank order in the two studies.

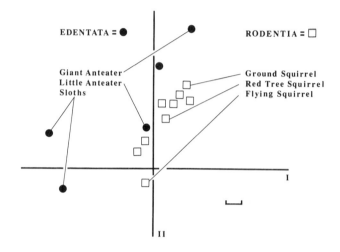

Figure 32. Bivariate plot of the positions in the first two canonical axes of selected genera of arboreal mammals. The marker represents one standard deviation unit. The separations produced among certain edentates and rodents parallel those found among the primates.

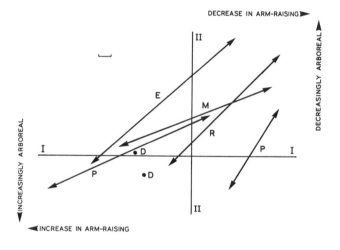

Figure 33. Bivariate plot of the disposition of a large number of arboreal members of different mammalian groups in canonical axes one and two. In each case, the general positions of the genera are indicated by the arrows. E = edentates, M = marsupials, R = rodents, P = primates, D = two isolated groups of dermopterans. The marker represents a single standard deviation unit. The genera are so disposed that differences in axis one relate to arm-raising and arm-hanging contrasts, and those in axis two associate with terrestrial arboreal differences.

the various nonprimate mammals. We may expect a number of possibilities to be the case a priori: that the kangaroo may appear similar to man because it does not utilize the forelimb for locomotion, that the otter may prove similar because it does utilize its forelimb for extensive manipulation, or even that a variety of small mammals may resemble man because they both do not use the forelimb much for locomotion and do use it for manipulation (e.g., gerbils and jerboas). Such possibilities spring to mind readily, yet if they proved to be correct, their very naïveté would suggest that the findings were spurious. However, the totally unforeseen results are shown in the bivariate plot of canonical axes one and three (figure 34)

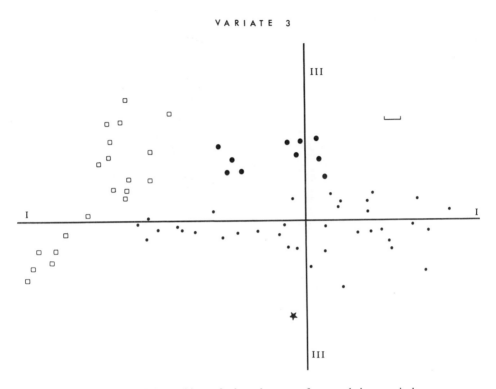

Figure 34. Bivariate plot of the positions of arboreal genera of mammals in canonical axes one and three. The marker indicates a single standard deviation unit. Small dots represent regular arboreal genera. Large dots represent the position of the various gliding mammals. Open squares represent the positions of some genera of bats. The star indicates the position of man.

and can be seen even more clearly when a three-dimensional model of axes one, two, and three is constructed (plate 2).

Thus, as in the primates, mammals which are only arboreal are not separated by canonical axis three. There are, however, some ten genera which are separated by this axis in a direction which is the opposite to that of man. All of these have a locomotor peculiarity in common: they are all gliding mammals. They include two types of "flying" lemur (a small gliding mammal of Southeast Asia belonging to its own Order, Dermoptera), three species of "flying" phalangers (Australian gliding marsupials), three species of "flying" squirrels (two gliding squirrels, and one species, *Anomalurus*, which also glides but is less obviously a squirrel), together with two species of giant tree squirrels which, though not normally recognized as gliding mammals, are capable of employing aerodynamic factors during their long leaps, especially on landing. Among the forms examined, there are no gliding mammals that do not show this trend. The separation takes on further significance when it is realized that some of the genera of bats are the only other mammals separated in this way by canonical axis three.

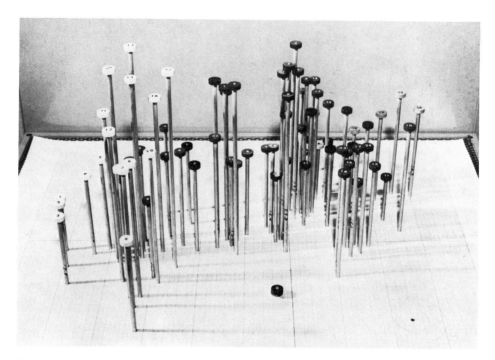

Plate 2. A photograph of a three-dimensional model of the positions of various arboreal mammalian genera in canonical axes 1 (width), 2 (depth), and 3 (height). The model shows well the generally intermediate positions of most arboreal forms, the high positions of the giant tree squirrels together with the truly gliding mammals, and the positions of the various genera of the bats (white). The unique position of man is also obvious.

However, these separations, interesting as they may be, do not shed any particular light upon the position of man. For while the gliding species lie in more positive positions in axis three, man remains unique in his more negative location.

The finding supplies further confirmation however, that the canonical analysis has in some way or other picked out aspects of the shape of the scapula which are related to its widely variant locomotor functions. After such intensive testing, we must assume that these biologically meaningful trends genuinely exist. Is it possible to test this idea in any other way? Certainly it would be more helpful to biologically oriented minds if we could "see" an actual relationship between the numerical separations achieved by the canonical analysis and the shapes of the different scapulae.

A Return to the Original Dimensions

Yet another test of possible biological meaning associated with mathematically derived functions has been made by returning to the original dimensions. The attempt has been made to see if the particular combination of original measurements in the different canonical axes makes morphological sense. In fact, we find that the combinations, rather than being merely arbitrarily arranged, do reflect obvious overall shape changes in the bones. This information can be obtained by deriving the relative contributions of the various original variables to the newly defined canonical axes. Such contributions differ somewhat, of course, from group to group, but the principal contrasts are those shown in tables 1, 2, and 3. These tables make it clear that the same partitioning of the variables is responsible for the positions of many different genera within the canonical space, both in the study on the primates and when information from various nonprimate mammals is interpolated.

Of further interest is the nature of the major contributing factors themselves. Thus the actual variables can be examined to see whether they measure coherent elements of the shapes or whether they are merely "inexplicable" conglomerations of variables. The shape changes are in fact reasonably coherent, and they can be best demonstrated as sweeping deformations when the appropriate shapes are examined in Cartesian grids. Furthermore the very nature of the deformations makes considerable sense in biomechanical terms, being easily referable intuitively to the biomechanical situations. This has since been confirmed by independent experimental stress analysis techniques capable of assessing the mechanical efficiencies of the shapes (see chapter 6).

Thus the contributions of the original dimensions to the first canonical variate are such that, taken in combination, they appear to be associated with a cranio-lateral twisting of the scapula and clavicle (table 1 and figure 35). This seems to be a feature that is related mechanically to the efficiency of the shoulder as a suspensory mechanism. Thus the contrast responsible for the separation of animals which do not easily and efficiently raise their arms in suspension, as compared with those that do, is the contrast between (a) those in which the scapular movement is principally a cranio-caudal rocking to allow movements mainly in the lower quadrant, and (b) those in which the whole shoulder girdle is twisted so that the shoulder joint is already cranially oriented before arm-raising occurs, thus presumably increasing the range and power of movements in the upper quadrant. Hence the correlation previously determined (that in this canonical variate the forms are distributed according to an intuitive guess about the extent to which

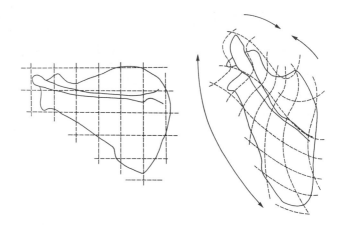

Figure 35. Two shoulder girdles (*Papio*, left, and *Gorilla*, right) which differ only in the first canonical axis. The deformation of the Cartesian coordinates for the gorilla demonstrates the primary difference in shape that exists. It is a cranio-lateral twist of the girdle of the gorilla as compared to that of the baboon.

TABLE 1   Contribution of Individual Dimensions to the Separation Achieved by the First Canonical Variate in the Arboreal Mammals.

Group	Dimension								
	a	b	c	d	e	f	g	h	i
Marsupialia	2	3			1			2	2
Dermoptera	2	3			1	3		3	2
Primates	2	3			1	2		2	2
Edentata	2	3			1	3			2
Rodentia		3			1	2		2	2
Carnivora	2	3			1	2			2

1 = contribution of more than 25 percent
2 = contribution of 15–25 percent
3 = contribution of 5–15 percent

the shoulder is raised and bears, or is capable of bearing, tensile forces) can be seen to rest upon an appropriate combination of original measurements.

Again, for the second canonical variate, the contributions of the original measurements are not meaningless or random; each seems to be associated with some facet or other of shoulder shape that is expressed in terms of a medio-lateral increase in the length of the clavicle together with an equivalent decrease in scapular width (table 2 and figure 36). In other words, the contrast responsible for the separation of the different forms in this axis is between (a) those forms in which the clavicle is short, the scapula placed on the lateral body wall and hence long in its gleno-vertebral dimension (extremes include terrestrial forms) and (b) other species in which the clavicle is relatively long, and the scapula placed dorsally on the trunk and is therefore short in its gleno-vertebral dimension (extremes include treetop-living forms). These latter features, as compared with the former, are those related to a mobile shoulder that is more efficient within a three-dimensional arboreal environment, rather than those related to a less mobile shoulder that is more efficient within the simpler cranio-caudal movement pre-empted by a two-dimensional substrate.

The third canonical variate is less easy to understand because there is only one form (man) among the primates that exhibits differences in this axis. There are no parallels here to guide us. And even when we trespass among mammals, although fascinating information about the third axis is found (for instance gliding and flying forms are separated by this axis in a positive direction) and examination of contributions of individual original measurements to this third axis seems to make considerable functional sense (table 3 and figure 37), there exist no parallels to the separation of man.

TABLE 2   Contribution of Individual Dimensions to the Separation Achieved by the Second Canonical Variate in the Arboreal Mammals.

Group	Dimension								
	a	b	c	d	e	f	g	h	i
Marsupialia			1	−2	−2	2		−3	1
Dermoptera	3		1	−2	−2	3		−3	1
Primates	3		2	−2	−3	3			1
Edentata	2		2	−2	−3	2			1
Rodentia	3		2	−3	−3	2			1

1 = contribution of more than 25 percent
2 = contribution of 15–25 percent
3 = contribution of 5–15 percent

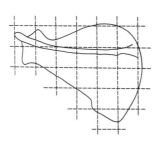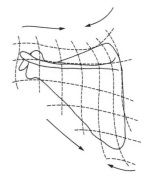

Figure 36. Two shoulder girdles (*Papio*, left, and *Pithecia*, right) which differ primarily in the second canonical axis. The deformation of the Cartesian coordinates in this case demonstrates that the main difference relates to a mediolateral compression of the girdle of the saki monkey as compared with that of the baboon.

Conclusions on the Primate Shoulder

There can be little reasonable doubt that in these particular studies mathematical variates have truly revealed biological facets of the shapes involved. Presumably this can only have happened if the mathematical system is modeling, by chance, what has occurred during the macroevolution of the group. Because the mathematical method has picked out a small number of aspects of the shapes (these are by definition mathematically independent), this suggests that the biological situation may also have been produced by the operation of a small number of biological factors of one kind or another. However, it does not necessarily follow that the biological factors need to be independent; it is likely that for many genera there is considerable association between the biological factors, though for others this seems to be absent. This is implied (a) by the generally linear form of the distribution of genera produced by the canonical analysis and (b) in the fact that there are sufficient genera that do not at all conform to such a pattern.

66   Form and Pattern in Human Evolution

TABLE 3   Contribution of Individual Dimensions to the
                           Separation Achieved by the Third Canonical
                           Variate in the Gliding and Flying Mammals
                           and in Man.

	Dimension								
Group	a	b	c	d	e	f	g	h	i
Marsupialia	3	−2	−2		1	1	2	3	−3
Dermoptera	−3	−3	−3		1	1	2	3	−3
Rodentia		−3			1	1	2	3	−3
Chiroptera	−3	1			3	1	2		
Man		*				1	2		*

1 = contribution of more than 25 percent
2 = contribution of 15–25 percent
3 = contribution of 5–15 percent
* = contribution of below but close to 5 percent

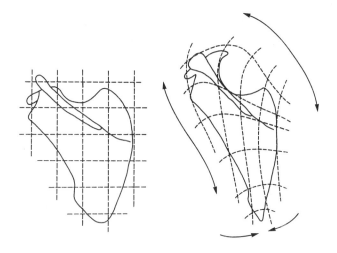

Figure 37. Two shoulder girdles (*Rhinopithecus*, left, and *Galeopithecus*, right) which differ primarily in the third canonical axis. The deformation of the Cartesian coordinates in this case demonstrates that the main difference is a cranio-lateral stretching of the girdle of the "flying lemur" as compared with that of the langur.

   Subsequent tests relating to these biological speculations from the canonical analysis include the use of independent mathematical and biomechanical methods. Both of these are described in later chapters and both confirm the above hypotheses. The robusticity of this information over many different techniques is the prime source of the idea that there is genuine biological meaning in these analyses. They are interpretations which cannot be discerned except by such studies (Oxnard, 1970).

For instance, this information is foreshadowed only vaguely by the dissectional studies of the muscles and bones (Ashton and Oxnard, 1963) or indeed by the inspection and simple analyses of the original metrical study of the shoulder girdle (Ashton and Oxnard, 1964) that form the original background to these multivariate studies.

The recent restricted study of the shoulder of the chimpanzee and man (Ziegler, 1964), the intensive functional studies of Inman, Saunders, and Abbott (1944), the very excellent dissectional and osteological study of the primate shoulder by Miller (1932), and the thorough metrical study of scapular shape by Frey (1923), do not provide suggestions about the evolution of the shoulder in primates other than that there is increasing mobility of that joint as one passes from prosimians, through monkeys and apes, to the epitome of mobility as seen in the human shoulder.

In contrast, the data contained in the individual canonical axes are totally novel and capable of providing, as we shall see, new insights into the mode of evolution of the free hand of man (chapter 7).

A Preliminary Study of the Primate Hip

A similar type of analysis has been carried out on a series of dimensions of the pelvis in extant primates. This is an augmented set of the data originally described in a univariate manner by Oxnard (1966), Zuckerman (1966), and Zuckerman, Ashton, Oxnard, and Spence (1967). In this case nine variables are defined on each of 441 innominate bones in forty-one genera of extant primate. Univariate analysis is clearly able to define the uniqueness of the human pelvis. Differentiation between nonhuman genera is, however, not nearly so clear.

The canonical analysis of this data (reviewed by Zuckerman, 1970, and to be described fully by Zuckerman, Ashton, Flinn, Oxnard and Spence) shows, first, among the many nonhuman primates, a marked set of separations that appear to make good sense in the light of the different functions of the pelvis within the different postures and modes of locomotion that exist. For instance, slow, clinging lorisines and the hopping galagines are separated both from each other and from all other groups. The relatively primitive, more pronograde New World monkeys and the rather similar lemuroid prosimians are grouped together. The more orthograde prosimians (Indriids), New World monkeys (Atelines and *Alouatta*), and all Old World forms (also more orthograde) make up yet other groups. These separations are achieved primarily by a single canonical axis. The second canonical axis affords a number of further separations, though its major effect is the unique separation of man from all other primates. A further two axes suffice to complete the separation of almost all the genera within these major categories from one another.

We are aware that these comparisons involve animals of very different shapes and sizes. For instance, statistically significant differences are not at issue here; it is obvious that gorillas are different from marmosets. What is required is characterization of the nature of the difference between these many forms. And because one of the gross distinctions is overall bodily size, a second series of analyses has been carried out utilizing data that have been transformed according to regression analyses performed against individual components of the pelvic dimensions.

In this case the general impression given by examination of individual angles and indices is that, though they differ in detail, they are similar in the overall picture of pelvic dimensions that they demonstrate. When canonical analyses are performed on this new set of regression-adjusted data, this general impression is clearly confirmed: the same sets of individual genera fall out in similar positions on the first (and major discriminating) axis.

However, there is a difference in the second axis that does suggest that a size factor may have been present in the first analysis that is removed in the second. Thus in the first analysis, canonical axis two performs essentially a separation (a) of man from everything else, and (b) of lorisine from galagine prosimians. In the new analysis, the separation of man from everything else is still present. But the lorisid genera (which are all similarly sized animals) are not now separated from one another by this axis; in contrast, the apes (differently sized) are now separated from the other Old World forms. The nature of this latter separation is of interest; for it is in a direction towards man: from genus to genus the separation is such that the smallest (*Hylobates*) remains nearest the Old World monkeys, the next genus is the somewhat larger chimpanzee, the next is the fairly large orangutan, and the nearest is the gorilla.

Thus it seems possible that the effect of regression adjusting the data (i.e., removing some elements of the size differences among the forms) has been to close the gap between man and the great apes and to widen the gap between smaller and larger forms. In other words, some considerable portion of the difference between the larger and smaller forms may be hidden by size differentials. As those differences in shape happen to be in a direction similar to that in which man is separated, as they concern man's nearest living relatives, and as they also effect, as we shall see later, man's presumed even closer fossil relative, *Australopithecus*, this may be an important finding.

4 Some Simple Testing Procedures

Introduction

While describing, in previous chapters, some applications of different investigative methods to studies of primate morphology, my self-imposed brief has been to attempt to render the techniques understandable to one without an extensive specialist background. However, in so doing, I have not mentioned various preliminary and pilot studies that are carried out before one launches into full-scale analyses. These preliminary investigations range from procedures as simple as inspecting a column of figures, through to the use of the final techniques upon subsets of what will eventually be the complete data. In original publications such preliminary studies may be mentioned, but they are often dismissed in a sentence or two that may give the reader little inkling of what actually has been done. Some studies may not even reach print at all. However, in a review description such as this, it seems pertinent to provide an outline of such preliminary and pilot studies. Most of the methods about to be discussed are not new; numbers of authors have produced excellent reviews developing and displaying them (e.g., Tukey, 1962; Healy, 1968 a,b). In this sense, what follows is not original; we have always leaned heavily upon expert statistical collaboration and consultation in these matters. What may be of value is the particular manner in which we use such preliminary work, the specific way in which we have linked the various studies. Tests of this nature (not necessarily this particular series of tests) should always be carried out (if not reported) in original research in this field.

In particular, I have been led to include materials on this topic as the result of the recent appearance of publications which have pointed, and rightly so, to some of the tests that should be made upon data, but which have implied that in anthropology, at least, workers are not aware of these problems. The implication as applied to some researches may be correct; but knowledge that such tests should be applied and accounted for is very old; it represents, in fact, a considerable proportion of the history of anthropology, and a number of earlier, excellent papers from the pens of biological statisticians are available that spell these matters out in detail. Certainly in our own work we have been using these tests for many years.

To what sort of preliminary work am I referring? It is possible to group different tests under a series of headings for the purposes of providing a descriptive framework, though such a framework is somewhat artificial, for a number of the methods are inextricably mixed in more than one category.

The various tests may be related to the gathering and sifting of data:

1. In the gathering of data, it is clear that the nature of the collecting process (the technique of observation or measurement) is open to various errors; such processes must therefore be examined so as to discover and eliminate or allow for such bias.

2. There may be peculiarities of data that are related to the fact that they are derived from biological objects of one kind or another. Features that are "biological" and that we ought to know about may have been introduced into data.

3. There may be facets of data that are due to the nature of the data themselves; one almost trivial (but see Healy, 1952) example of this is that data will clearly vary if taken in different units on separate occasions.

Once data have been gathered it is usual to apply a second set of techniques to operate upon them:

4. These techniques are those of data analysis and should also be tested as one would test methods of data collection.

5. Yet another set of tests that may be applied relates to the actual results obtained from analyses: can they be investigated in order to determine if they are "reasonable"? This is not the same as testing the method of analysis; for a well-tested analytical method may yet provide faulty results because the particular data may violate some necessary statistical assumption. This may sometimes be easier to determine by performing several analytical test runs on the data, than by preliminary examination of the data themselves.

6. Finally, from the results of analyses, speculations may result. These speculations, in turn, should be subject to testing; and, though some such tests may be looked upon as the next hypothesis to be tested, others may relate to smaller aspects of the problem that should be investigated as part of the current researches.

This second group of tests (4, 5, and 6) relates to the actual analytical methods themselves. Such tests may be fairly sophisticated and are described in chapters 3, 5, and 7. But the first group (1, 2, and 3) is also important and is relatively easy to carry out. These tests are treated below.

The Technique of Data Collection

When we make a measurement or observation, we use an instrument of some kind or other; the process is a technique which should be tested and, if necessary, calibrated. If the instrument is simply the naked eye, there is little we can do to make appropriate tests. Such tests which can be made devolve mainly upon psychological elements. They relate to the nature of observer bias; and they include such problems as observer preference and observer limitations. Observer preference is well known; in one recent botanical study, analysis established that one of the main trends within the data related to the differential preference of the different student observers. The question of observer limitations is beautifully displayed by Gibbs (quoted by Healy, 1952), who finds that the frequency of misreadings on different types of aircraft altimeters varies from as little as 0.5 percent when figures are displayed in an aperture, through 10.9 percent when displayed by a moving pointer on a circular dial and as much as 35.5 percent if the pointer moves on a vertical scale. Although we have not ourselves carried out tests of this general nature, we are always most careful to check interobserver bias when, as is sometimes necessary, two people are responsible for gathering the data.

When, however, our techniques utilize instruments other than the human eye, they are always tested by the method of comparing the variance within replicates (i.e., variance introduced by the technique) with that between specimens (i.e., including variance which genuinely exists). This is done to determine if the technique is capable of distinguishing differences over and above those produced by its very use. As this may seem obvious to the reader, he may question if it is necessary. The answer probably is that it depends upon what is being done. For instance, in making measurements of the overall length of some bone with calipers, a test of this nature is marginally required. But with exactly the same pair of calipers, the measurement of some muscular marking or articular facet, for example, must be very seriously tested. This is necessary because we wish to test not only the accuracy of the instrument, but of the whole procedure. And though in the previous case almost all workers can agree upon what constitutes the maximum length of a bone, there may be a much greater degree of *legitimate* argument about the limits of a muscular marking, the edges of an articular facet, or even precisely how to hold the bone while the measurement is being taken. One still comes across the occasional student who feels that what is being tested is the accuracy of the calipers and who does not realize that the technique comprises such things as the orientation of the specimen before measurement, the decision of what to measure, the actual use of the calipers, the method of marking the result, and even the elimination of memory as an element in data collection.

These remarks apply whether the technique is a simple measurement or a complex phenomenon like photoelastic analysis. The precise way in which a test may be made varies with the technique: in osteometry, as described above, a test may be rather a simple matter. In photoelastic analysis, testing the technique may involve a whole series of experiments concerning such things as effects of ambient temperature changes, effects of different samples of photoelastic plastics, effects of difficulties in loading specimens, tests for time-edge effects, tests for strain-creep, tests for the introduction of edge stresses, and so on. This list is given not to impress but to indicate that testing may be a major part of an investigation. Indeed, in certain types of study (photoelastic analysis, for example) it may even be necessary to make tests during investigations to confirm that conditions have not changed or, if they have, to know how they have changed so that allowance can be made for them in the analyses of the results.

My own experience is that some workers never test their methods, and that although most tests confirm the validity of a technique, one is continually surprised by techniques which, on testing, prove to be unable to distinguish real biological differences. This has indeed proved evident once in my own work, in which a technique was adequate for examining differences among certain parameters of the shoulder girdle in mammals; but when the study was extended to further specimens, the tests showed that the technique was insufficiently sensitive for measuring specimens smaller than those previously encountered (Oxnard, 1968a).

Biological Facets of Data

In the preliminary scanning of data it is necessary to make explicit examination for perturbations that may result from the biology of the situation. The most obvious way in which data may be so affected relates to the nature of its frequency distribution, that is, the distribution of the values of each specific variable when examined both within and between the fundamental groups[1] of specimens used in the studies.

Now in an entire study it is highly likely that there will be many fundamental groups which are so small that the nature of the distributions of data within them cannot be investigated. For instance, in the studies on the shoulder girdle there are nine groups which comprise as few as four specimens and even three groups containing single specimens only. Such limitations of this data can scarcely be improved by further data collection; there just are not the specimens of genera such as the aye-aye, the woolly spider monkey, or the pygmy chimpanzee available to us to enlarge these groups. This does not imply that the information given by these small samples is useless; on the contrary, such information may be of special importance (rare specimens are often also those that are aberrant and of significance in the evolution of a set of organisms).

We must therefore do the best that we can with the materials available. One way of doing this is to examine frequency distributions of data within a smaller number of the fundamental groups chosen as test pieces. In making such choices, it may be necessary to utilize certain groups simply because a large enough number of specimens can be assembled in one way or another for the examination of frequency distributions. It is also necessary, however, to choose some test groups in such a way as to magnify as greatly as possible the likely perturbations that may exist within part or even the whole of the data set. For instance, in the studies of the primate shoulder, the following genera were chosen for such tests: *Homo, Gorilla, Pongo, Pan, Cercopithecus mona,* and *Leontocebus;* in the studies of the mammalian shoulder the following additional genera were also chosen: *Erinaceus, Petaurus,* and *Calliosciurus,* together with *Tadarida, Desmodus,* and *Sphaeronycteris.* The reasons why these different genera were chosen are related to the different problems that may be encountered.

First, it is clear that in studying such a wide range of animals one major difference between them is that of overall bodily dimensions. Within primates these range from the extremely large (*Gorilla* and *Pongo*) through intermediate forms such as *Cercopithecus,* to the very small *Leontocebus.* When a series of mammals is also investigated, a further decrease in size is noted among the specimens and hence a group of small mammals roughly equivalent to *Leontocebus* was examined additionally (*Erinaceus, Petaurus,* and *Calliosciurus*). As a final check the smallest of all (the 3 genera of bats) were included. Such studies are capable of examining two possible effects: one includes the tests of technique that have already been discussed; a technique that is valid for the gorilla could easily fail when applied to the vampire bat (*Desmodus*); a second effect relates to data dis-

---

1. By "fundamental groups" we mean the arranged groups being investigated. This assumes for the moment that known groups (genera, species, subspecies, etc.) are being examined, each comprising several specimens, rather than a totality of specimens belonging to a single group.

tributions that differ simply because the specimens are of different sizes. The real variances of actual measurements may be much greater for large forms than for small. Even when indices are used (which is a simple attempt to circumvent this effect), it is possible that the variances of a given index may differ among the groups. Hence, because we can scarcely test this on the full body of specimens, it seems important, at the very least, to examine the situation within representatives of different size classes.

Secondly, a whole series of problems of distribution of data relates to our concept of fundamental group. Thus the term *fundamental group* has been taken by some workers to mean *that subset of the specimens, of whose internal structure one has not yet become aware.* Such a definition may be useful in examining an entirely new universe. But in all our studies a great deal is already known about the different animals; the particular groups used are forced upon us by the paucity of specimens. Had we been interested, for instance, in skull structure, it would have been clearly possible to examine the data in terms of species, or even subspecies or local populations. Because we are interested in the postcranial skeleton, the collecting habits of mammalogists limit us to samples at the generic level. Nevertheless we know that such subgroupings do exist and that they may affect distributions of the data. Accordingly our tests include genera containing subgroups in order to assess the extent to which other genera, for which we do not have representatives of subgroups, may be affected. And though to some considerable degree the particular genera chosen are imposed upon us by the available samples, at the same time this is not entirely the case; we may struggle to obtain extensive samples of particular genera simply because we know that they are extreme in this way.

One obvious subset of a living group that must be examined is sex. Sexual dimorphism may be of fantastic proportions, for instance, in arthropods. But even in mammals it is a truism that differences between sexes may be enormous, notwithstanding the fact that such differences may be very highly correlated with one another, being dependent upon a rather small number of physiological, genetic, or other parameters. Hence, our tests for sex have always included the gorilla and orangutan because these are the primates that display the greatest sexual differences. In addition, we have also studied sex differences in man even though human sexual dimorphism appears small. This is almost the total extent of testing that has been possible with the primate collections available to us. We have never made such tests upon material unless sexing was carried out in the field. It would seem possible to utilize museum specimens for such tests in species such as baboons, where one might think that sexing can be readily based upon the characters of the specimens. But this may introduce circularities in argument; the possible fallacy here has been highlighted by an amusing example brought to my notice by a graduate student who desired to know the sexes of a sample of pig forms (*Babirussa*). Among the specimens available to her at the commencement of her study were some with large tusks and some with small and these were tentatively determined as males and females. Only later, as her knowledge of these pigs increased, did she come to recognize that certified females have virtually no tusks at all—her entire original sample was (presumably) male.

The importance of carrying out tests upon sexed materials is clear. For actual measurements, differences between sexes will sometimes be far greater than those between the fundamental groups of interest—in our case, genera. By the judicious use of indices we have been able to "reduce" sexual differences in the dimensions of the shoulder; presumably such differences as exist in the raw measurements are due in large part to relative sizes and are eliminated by the use of the index. However, there may well exist elements of sexual dimorphism that are not related to size; this may be much less obvious to the eye, and may occur in groups of animals other than those we have examined; in other words, our tests are by no means foolproof. Our actual findings are that in both the shoulder and pelvic regions, ratio differences between sexes are mostly insignificant. Those differences which are found to be significant are small compared to those which we are interested in studying, namely, the differences between genera. Accordingly, the data are pooled in those genera where sex is known; comparisons may then be made with those groups (the great majority) for which sex is not known.

Pooling of data in this way may produce bimodal or skewed distributions in the groups where sex is known; similar perturbations from normality exist for the great bulk of the data where sex is unknown.[2] However, with adequate testing some idea of the degree of such perturbations may be obtained. When using indices and angles, this has proven small, even in genera markedly dimorphic. Given the much larger differences between genera that we are attempting to examine, and given the robust nature of many statistical methods, we may then proceed to the main study with some confidence.

Everything that has been written above about sex subgroups of the data may be applied to the various other possible subgroups. Infraspecific subgroupings may be considerably more extensive for some genera than others. Perhaps one of the most marked examples in this respect is man; accordingly, in all our studies, man has been used for testing problems of this kind. It has always proved possible, because of the extensive collections of human material that exist, to gather sufficient samples of at least some of the subgroups of man to perform appropriate tests. Of course, one is keenly aware of the problems of making such tests in man; the reality of some of the groups that we may examine is speculative, both in biological terms and in terms of definition of museum materials.

We have been talking above about genera represented by a single extant species. An equivalent test that may be attempted is comparing species within genera. Again, deficiencies of materials that have been available to us from various museums have limited the number of genera that could be tested for this particular effect. However, in the studies on the shoulder such tests have been made on *Cercopithecus, Macaca, Cercocebus,* and *Presbytis.*

---

2. Tests for peculiar data distributions are described later in this chapter.

In these cases our results are not the same as when testing sex or human racial groups. For between the different species of these genera we have found differences that are significant. In one case (*Presbytis*) the differences are both significant and fairly large; the mainly ground-living langurs (only those of the species *entellus* were included in the test) are considerably different from the predominantly tree-living forms (again confined, for purposes of the test, to the species *obscurus*). Accordingly, because we have a reasonable number of specimens, this genus is subdivided in the main analyses. There is considerable external evidence (e.g., about the function of the shoulder) that makes this seem a sensible procedure.

For the other genera examined in this way some significant differences are also found, though they are of considerably smaller degree compared with differences between genera. Nevertheless the nature of the interspecific differences is such that they appear to relate to known differences in the extent to which the species are terrestrial or arboreal. Thus the terrestrial species differ from the arboreal species for *Cercocebus, Macaca,* and *Cercopithecus,* and, in this last case, an intermediate species (that is, intermediate in terms of the variables measured on the shoulder girdle), may also be defined as intermediate on habitat grounds. These particular genera, however, are retained as pooled groups. This is done partly because the degrees of the differences are small as compared with those between genera, and partly because we fear that prejudging the issue of the final analysis by the use of these findings may falsely channel the results of the full analysis. In the case of the langurs, the habitat differences are so obvious and well known, and the morphological differences are so great, that we do not feel that the procedure has entailed loss of objectivity. In any case, to split a single genus in a forty-genus study would seem unlikely to bias the eventual results; the effect of splitting half a dozen genera might be something else.

It is nevertheless clear that we have been less than satisfactory here. There are a number of other primate genera in which functional differences between species within genera may exist (for instance, there is some evidence that the different species of howler monkey may move differently in different environments) and in which we would like to be able to examine existing intrageneric variation. Paucity of material prevents such study.

Other subgroupings of data that may exist and for which one should attempt to test relate to such features as the existence of age subgroups, of subgroups of data from captive animals as compared to field-collected specimens, of subgroups of pathological specimens, and so on. In our osteological studies we have not had sufficient material to make such tests. Accordingly, young, captive, and pathological specimens have been eliminated from our series except in the case where rejection of immature material would have left us with no information at all about certain important but rare genera. Even here such data are viewed in a circumspect manner with a smaller degree of confidence placed upon them.

In soft tissue studies however, we have more often been forced to accept markedly immature material. Preserved material, especially of adult specimens of orangutans and gorillas, is more difficult to procure than osteological material. This is certainly the case in our studies on the musculature of the primate shoulder; it is even so for the more extensive soft tissue studies carried out by Tuttle (1969, 1972) on the musculature of the extremities. Accordingly, we have felt it essential to attempt to test in some way the effects of age upon ratios of muscle weights.

This could be done, for example, by dissecting five or six specimens in each of five or six age groups of a given genus. Such a study, similar to a test of technique, would be extraordinarily time consuming when applied to dissections. Consequently tests were applied to the very extensive data already available from the series of human subject dissected by Theile in 1884. Theile's data need to be examined with care, and much is rejected because pathological features (such as the nature of the disease process that caused death) interferes with the tests. Nevertheless, even with rigorous pruning, Theile's work is sufficiently extensive to gauge changes of muscle weight with age. We have found that, for the shoulder musculature, as long as the very earliest stages (up to milk dentition) are eliminated, then it is unlikely that the inclusion of some immature specimens markedly biases the data. Tuttle (1969) confirms this for the major part of his data, though he finds otherwise for the triceps surae in *Erythrocebus patas*.[3]

A second way of performing such a test is to attempt the same study on an easily obtained and easily dissected laboratory animal such as the rat. In this case other factors may upset the test; for instance, strains of laboratory rats are highly inbred and may show more obvious relative change with age because differences between animals of the same age may well be less. But a qualification like this plays against the investigator and is therefore probably not unsound. The results of this test also suggest that extensive changes in ratios of muscle weights with age do not occur except near birth.

In the final analysis, it is likely that the use of the muscle weight as a parameter of the morphology of a musculo-skeletal region is so crude as to be of little value. It must be left to the individual worker to determine whether the amount of information revealed by such a study is worth the effort. Should, however, it be judged that it *is* worth the effort, then the very crudity of the technique makes it all the more important to carry out an appropriate set of tests.

These, then, are some of the ways in which biological aspects of organisms may affect the nature of data that can be obtained from them. Many of these are obvious; some are a little more subtle and often are not explicitly tested for. The tests relate, in general, to specific null hypotheses: that the effect of sex on the data does not differ from zero, that the effect of age does not differ from zero, and so on. Such tests can be made using statistics such as t and χ^2. Though they are

3. But Tuttle believes that this could be related to a functional difference between adult males and others in terms of locomotor behavior that might be associated with this particular muscle. However, his method of testing (multiple t tests) does not rule out the possibility that his finding is spurious.

easy to compute, the nature of research is often such that there may be a tediously large number of them to handle.[4] Today this is no problem; often it may be a simple matter to have these tests done as preliminary calculations before undertaking a major analysis on an electronic computer. Sometimes, however, one may wish to have all or a part of this information before deciding on the nature of major analyses, sometimes even before collecting all the data. And although fifteen years ago a battery of tests like this might have occupied weeks or months of work for a single person on a mechanical calculating machine, advances in electronic desk calculating devices, capable of using short programs and of storing numbers of intermediate results, can help enormously. They allow many more tests to be performed with far fewer errors.

Problems Inherent in the Nature of the Data Themselves

The tests of the previous section have all related to specific items that are biologically identifiable and that may be expected to result in data which, for a given fundamental group, are not normal in distribution.

But it is sometimes forgotten that the data may be nonnormal for reasons that have little to do with the biology of the situation but rather are related to the logical nature of what is being done. Such problems may loom rather large, and their effects may possibly be considered in two ways. First, one can ask what possible factors in data may produce peculiar frequency distributions. Secondly, one may examine the frequency distributions that exist and see how greatly they vary and to what extent something must be done to take account of this variation.

How may the data themselves affect their own frequency distributions? When data are in the form of indices (or angles, ratios, percentages, etc.), problems may occur. For instance, if there is no correlation between the two elements of the index, then the significant figures of the index may be much less than those of either measurement; a considerable loss of sensitivity may occur. Presumably, however, this will not be the case, because one of the reasons for using an index is that both original measurements are elements of some overall structure that one wishes to characterize and both should be correlated to some considerable degree.

Of more importance are the problems that arise if one examines an index which is the result of dividing two counts (enumerations) one by the other. Here the final distribution of the index may be very different from the distributions of the original counts. Furthermore, whereas counts may have similar data intervals from one datum to the next, this will not be the case with the resulting index. Figure 38 reflects the sort of problem that may occur. And though this error is well described in numerous textbooks, one still sees examples of its propagation.

---

4. This is not to say that I am advocating a method of examining data such as the computation of all possible pairs of t tests. Significance in a small number of such tests when obtained from a very large battery cannot be judged as meaningful. Far better methods of analysis are available in such situations (e.g., multiple analysis of variance).

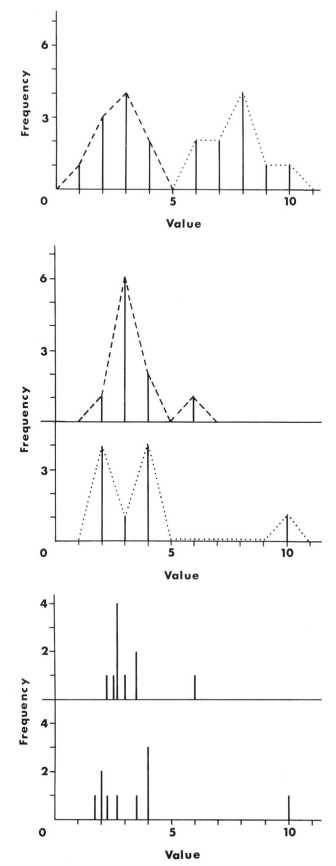

Figure 38. Simple problems with enumeration data. The first diagram shows values for two hypothetical counts expressed as frequency histograms. Although the numbers of specimens are very small, the figures have been arranged so that the distributions resemble normal.

The second diagram shows the two distributions that are obtained from the counts in the first figure when these are joined in a ratio. The upper distribution assumes complete correlation of the two counts (i.e., that the smallest of one is related to the smallest of the other, and so on). The lower distribution assumes random association of the two counts. In each case the histogram has been drawn using the nearest whole numbers for the parameter values. In both cases the distributions are nonnormal as far as can be judged visually.

The third diagram shows that the true values do not lie at equal parameter intervals. Not only are the distributions nonnormal, but also it is not justifiable to arrange the parameter values as whole numbers.

A more subtle form of this problem results from the use of a count as only one element of an index; it is clear that this too may result in a frequency distribution that is different from those of the original measurements (and usually different from normal). Even when both elements of the index are continuous and relatively normally distributed, the nature of the correlation between them may produce nonnormal distributions of the index; smaller perturbations from normality of the original measurements may, in particular cases, sum to large deviations from normality in the index.

Further problems may arise from the use of an index if it is in the particular form of a part being compared with a whole. In such a case, as can be seen from figure 39, though the frequency distribution of an index lying between 0.2 and 0.8 may appear relatively normal, gross variations from normality may occur as soon as measures of central tendency approach the limits 0 and 1. As the diagram shows, normal curves may become replaced by curves that are heavily skewed, or perhaps that may resemble other types of distributions such as the Poisson.

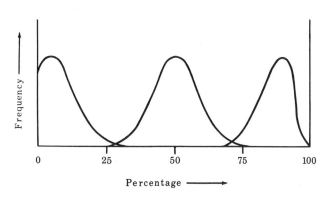

Figure 39. Possible frequency distributions for ratios which consist of a part divided by the whole.

On an even simpler level, data may be misrepresented if the worker is not aware of some problems relating to secondary manipulations. For instance, the first diagram of figure 40 shows data that are skewed. The second diagram of the sequence shows that increasing the grouping interval (a procedure that may well be justified) can, if it is done incorrectly, hide the skewness. The third diagram in which the grouping interval is increased correctly shows that the skewness is still present. Likewise, incorrect grouping may introduce skewness that does not truly exist.

These various problems should, of course, never occur. It merely requires that an individual worker pay attention to the logical nature of what he is doing when forming secondary indices from primary data of one kind or another. This is especially important when the primary data consist of dimensions which are other than measurements of length.

Notwithstanding carefulness in these regards, it is still possible that data distributions may depart from normal. To reveal this it is necessary to scan the data

Figure 40. Simple secondary manipulation of data. Data that are slightly skewed (first figure) may, if the grouping interval is changed, lose skewness (second figure), or retain it (third figure).

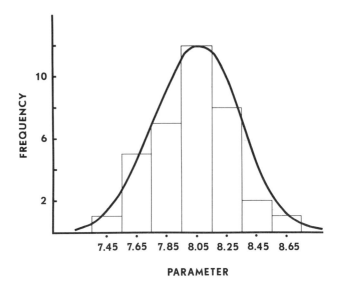

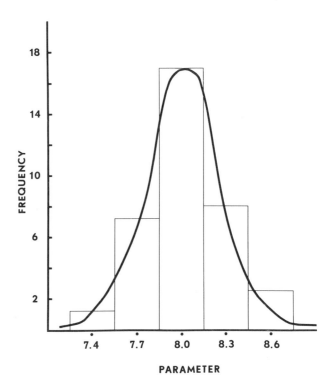

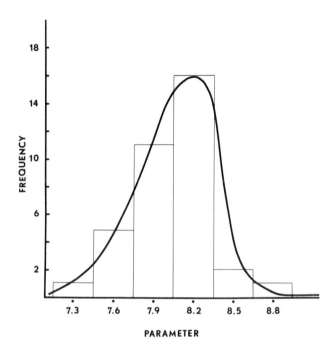

in some way or other. One particularly important factor that may apparently change the distribution of data is the presence of items that are mistakes of measurement, identification, recording, and so forth. Another may relate simply to the fact that the real population from which the sample is drawn may itself not be normal; in this case a nonnormal sample frequency distribution is *correctly* reflecting the population. In situations like these we are less interested in whether or not a distribution differs significantly from normal (the null hypothesis tested using the distribution of a statistic). Rather, we desire to know the actual data distributions so that we can identify the mismeasurement (for correction if possible) or characterize the nature of the nonnormality (so that we may apply an appropriate correcting transformation). Perhaps the best methods for these purposes are a variety of visual techniques. As with the simple tests previously described, these are quick and easy to use. They can be fairly rapidly done by hand; they can be very rapidly produced if one has access to simple plotting equipment such as is now available on the better electronic desk calculators.

One simple and effective technique is of course the histogram. Such perturbations as skewness and kurtosis (if gross) can be recognized. Coefficients to determine the degree of these can be calculated (Quenouille, 1952). However, if the number of data points is small, histograms may not clearly delimit milder degrees of skewness and kurtosis, and a previous example has shown how mismanagement of the histogram may hide or create features of this type. The combined use

Figure 41. The histogram and the "hanging rootogram." First diagram: histogram of parameter values and their fitted normal curve. Perturbations from normality are represented by the deviations of the histogram columns from the curved line. Second diagram: hanging rootogram of the same data. The number of observations is plotted on a square root scale. Perturbations from the fitted line are now distributed around the horizontal base line; visual assessment is much easier. After Healy (1968a).

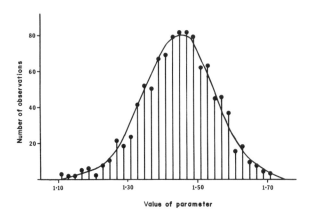

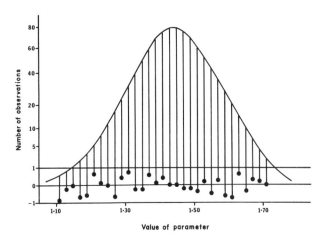

of the histogram and its fitted normal curve (figure 41) may point more easily to curiosities in the data. However, it is difficult to be certain about deviations from a curved line. A useful technique (suggested by Tukey and published by Healy, 1968a; see also Tukey, 1970) for clarifying the perturbations of a histogram from its fitted normal curve is the "hanging rootogram." Here all the vertical distances are replaced by their square roots, and rather than standing upon the base line, they are hung from the fitted normal curve. The feet of the histogram elements fluctuate about the base line; the square root transformation results in this fluctuation being associated with a constant deviation of plus or minus 0.5 units. The eye can thus be guided by the addition of limit lines one unit above and one below the base line. We would expect all but one or two of the histogram bases to fall between these limits (figure 41).[5]

5. The limitations of histograms are most evident when the total number of items falls below 50; here cumulative normal plotting may be more valuable.

A second rather useful searching tool in this context (and again an easily derived graphical method that can be scanned by eye) is normal plotting. This is done by plotting the individual sample values sorted into ascending order (figure 42). If this plotting is made utilizing arithmetic graph paper, the data may be arranged in an S-shaped curve. The fact that the curve is sigmoid suggests that the distribution is normal or near normal; clearly however, it is difficult to assess the degree to which an observed S-shaped curve differs from that of a particular expected (normal) S-shaped curve. If the process is carried out utilizing normal probability paper, this assessment is much easier to make. Normal probability paper is special graph paper on which the vertical scale has been stretched in such a way that the S-shaped cumulative normal curve is transformed graphically into a straight line. There are problems here, however. The y axis is graduated in percentages, and, as indicated by Healy (1968a), there has been considerable discussion on the best values to use for the members of a sample of N normal values. However, given that an appropriate scale is used, systematic departures from normality show up as departures from a straight line. Single extreme points can also be detected by this method (see figure 42). Normal plotting can be used even with quite small samples; the slope of a line on a normal plot is the reciprocal of the standard deviation.

Another departure from normality that may be of considerable importance is systematic change in variability. This is rather common and has already been touched upon in the discussion of tests performed upon specimens of widely differing overall bodily sizes. In this particular example, sections of the data with large mean values may have at the same time large variances. Such a situation usually requires a transformation of the scale of measurement, and simple graphical methods can be helpful in elucidating it.

When the data consist of a large number of groups, the choice of a variance stabilizing transformation may be guided by plotting the group standard deviations s against the group means $\bar{x}$. The resulting array of points can then be examined by fitting to them a linear regression line. If the regression line shows a marked and statistically significant positive slope, then the data may be transformed logarithmically and basic data (means and standard deviations) recomputed. Plots of standard deviations of the logged data against the corresponding generic means should then (if the transformation has been effective) show standard deviations which fluctuate randomly irrespective of the value of the mean (figure 43).

The use of the histogram and its derivatives, and of cumulative probability plots and their modifications, are ways of examining the distribution of individual variables one at a time. But it is a matter of experience that, for certain purposes, methods involving the simultaneous examination of two or more variables are more powerful for detecting maldistributions or outlying values. With a big sample it may be enough to plot a dot diagram of one measurement against another. As with a single variable however, this method (equivalent to plotting a

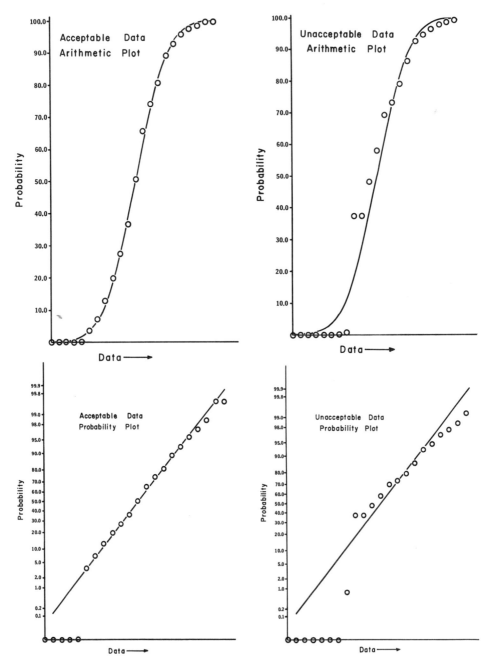

Figure 42. Probability plotting. Upper diagrams: probability plots on arithmetic paper of two samples of data. That on the left is known to be acceptably close to a normal (S-shaped) distribution (on the basis of statistical tests); that on the right is not. As with figure 41, because the data are fitted to a curved line it is difficult to say visually whether or not either given data set is acceptable. Lower diagrams: plots of the same sets of data on probability paper. Here the data should, if normal, fit a straight line. It is now easy to say, on visual grounds alone, that the data on the left are acceptable, those on the right are not.

Figure 43. The use of a transformation. Upper figure: plot of the mean against the standard deviation for a given parameter for a large number of different data groups. Although the relationship between mean and standard deviation is not visually marked, there is nevertheless a significantly positive regression. Lower figure: when the above data are logged and the basic statistical parameters recomputed, a plot of the means of the groups against the standard deviations now suggests that the association is random.

In this case a logarithmic transformation may be of value.

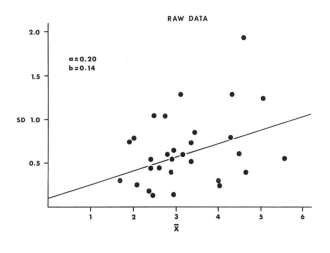

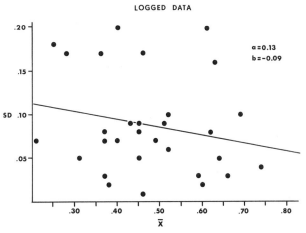

histogram) is less effective for small samples, and Healy (1968b) has suggested a technique analogous to univariate normal plotting, which can be applied to two or more variables simultaneously. This is shown in the next diagram (figure 44), where an appropriate distance measure[6] between two variables is plotted on normal probability paper. There is no reason why the method cannot be used for any number of variables. However, it is possible for the use of more than two variables to result in loss of power, because an outlying value in one variable may be balanced in terms of its "distance" by a number of centrally placed figures for the other variables.

6. A variety of measures of distance could be used. But if we wish to take into account the differing variances of the two samples and the degree of correlation between them, then an appropriate statistic would be D^2. The exponential distribution of the X^2 values associated with this can be normalized by using a square root or cube root transformation.

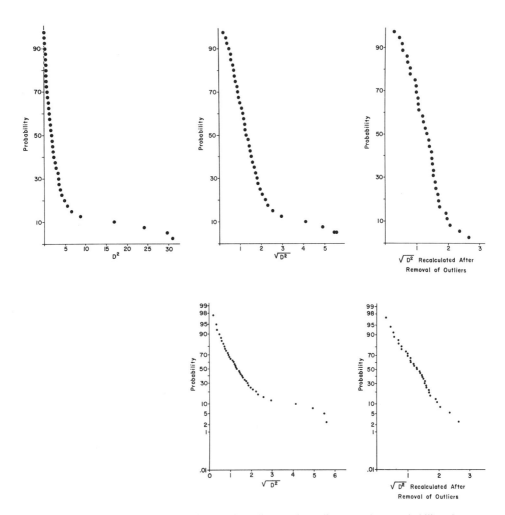

Figure 44. Multidimensional probability plotting. The top three diagrams show probability plots (on arithmetic paper) of generalized distances between two variables: left, plot of D^2 against probability; middle, plot of $\sqrt{D^2}$ against probability; right, a similar plot after removal of four outlying values and recalculations of $\sqrt{D^2}$ values. If the distributions are normal, such plots should approximate to S-shaped. The third curve is much closer to being S-shaped than is the second, so that we surmise that removal of the outlying values has converted the distribution to normal.

Again, however, a judgment as to degree of fit to a curved line is difficult to make visually. Replotting the second and third figures above on probability paper gives the probability plots shown in the bottom two diagrams. It is now quite clear that the full data set does not approximate to a straight line; the data set after removal of four outliers is, on visual inspection alone, remarkably close to a straight line.

This technique of multivariate normal plotting is a very powerful tool in the examination of data. It may help to show that attempts to summarize samples by way of means, standard deviations, and correlation coefficients may be totally misleading.

Conclusion

The description of these different preliminary statistical tools is not meant to supply a "best" way of checking these aspects of such studies. Indeed, with the many packaged computer programs that are now available, it may be relatively easy to attempt a more sophisticated set of tests (e.g., Wilk and Gnanadesikan, 1968). Perusal of many statistical texts will show a large number of such tests which in many instances may be more efficient than those that we have chosen to use (e.g., see Fisher, 1958; Siegel, 1956; Snedecor and Cochran, 1967; Bradley, 1968; Sokal and Rohlf, 1969). Because of initial limitations on our computing abilities (this work has been in progress since before computers were readily available), we have tended to rely on visual methods (e.g., Quenouille, 1952); yet today computational answers may be more easily obtained and certainly more quantitative in nature (e.g., see Preston, 1970). The value in presenting the above information rests in showing that it has been done for the current studies (an important strengthening factor in their interpretation), and in drawing the attention of the reader to the fact that something of this sort should always be done to validate similar studies.

5 Group Finding Procedures in Morphology

Some Limitations of Multivariate Methods

The methods of multivariate analysis are designed for the investigation of data which conform (or nearly so) to distributions that are Gaussian in nature. That is, for one- or two-dimensional sets, the arrangement of the variation within groups approximates to the familiar bell-shaped curves and surfaces. Appropriate hyperbell-shaped distributions are assumed for data in more than two dimensions. And it is certainly true that in many cases where extensive sets of biological data have been investigated, normal or nearly normal distributions are common. However, there is no particular reason why they should be ubiquitous, and, indeed, there is every indication that they are not.

Another requirement of many methods of multivariate statistics is that common variance and covariance structures should exist throughout the groups being examined. When a biologist views comparatively two species of animal, he is generally thinking about the means and spreads of the various features that he sees. He usually assumes that the degree of spread, as assessed by eye, is similar for each character and he often believes that the correlation (considered intuitively) between any two characters is approximately the same for each species. However, even at a nonmathematical level, biologists may be well aware that certain animals are different from the majority with respect to variation and correlation; for instance, Davis (1964) recognizes that the giant panda *Ailuropoda* possesses a series of apparently "ill-assorted disharmonies" that result in many of its morphological characters displaying far larger variation than is the case for the bears in general (pace those who believe that *Ailuropoda* is more closely allied with the Procyonids). In like manner, many primatologists know that, in comparisons involving the orangutan, the "grotesqueness" of this animal is seen as "excessive" variation within its morphology that is of an order different from that found in most other primates. Indeed in certain multivariate studies (Ashton, Healy, and Lipton, 1957) it was found necessary to treat the orangutan separately because of this.

The questions that we are asking when we use multivariate statistical methods are further restricting factors in their use. Thus factor analysis attempts to define structure within what is thought to be a *single group* of data (taking account of the variation and covariation within that group). Principal components analysis asks for that structure to be viewed in those mutually orthogonal (independent) directions that reveal the greatest structural dimensions. Generalized distance analysis attempts to elucidate the differences between what are known to be a *number of groups* of data (taking account of the variation and covariation within and between the groups). Canonical analysis investigates these distances in mutually orthogonal (independent) planes producing maximum separations among the groups in the multigroup space.

But in many cases with biological data, the questions that we are interested in asking do not relate to given groups. Rather, we are interested in knowing if there are any groups in the data and, if so, what are the structures within and the relationships between them. And we are also interested in knowing these answers even if the groups do not approximate to multivariate normal distributions. Groups may be sausage-shaped and even doughnut-shaped in three dimensions (hypersausages and hyperdoughnuts in more than three dimensions), and these are facts in which we are most interested.

Finally, and as a corollary, we may well be interested in defining groups in sets of data which are only locally stable. For instance, if, in utilizing multivariate normal techniques, it becomes necessary to enter new data, then the entire data set is reexamined and in consequence every point in the interpoint distance matrix takes up a new value. In biology there may well be no particular reason why the incorporation of new data at one point should affect other regions that are located some distance away. It is therefore of interest to utilize methods that have local stability and that do not involve the entire interpoint distance matrices.

For instance, let us consider a hypothetical data space such as is shown in the diagrams (figures 45, 46 and 47). These attempt to demonstrate, first, that the arrangement of specimens may be such that groups may exist which possess a variety of shapes and configurations and which are other than multivariate normal in nature, and, secondly, that these nonnormal groups may in fact exist in a data space even when the distributions themselves approximate to normal on each reference (measurement) axis. There are distinct possibilities of different types of data groups: one is related to high mutual similarity of members (aggregated groups); another is typified by sharpness of separation (segregated groups). (These two types of groups appear similar to what have been described as "homostats" and "segregates" in certain psychometric studies; see Cattell, 1968.)

On the one hand, the groups enclosed in figure 46 are those sets of specimens coalesced because they contain similarities (of the x's and y's); on the other hand, those groups enclosed in figure 47 are similarly isolated from others. The first type of group, the aggregate, is "real" for some purposes; for example, within

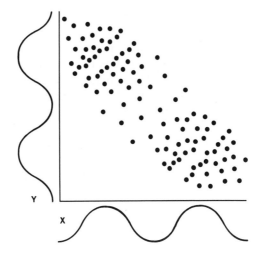

Figure 45. Bivariate plot (x against y) in which the distribution of each individual variable approaches normal.

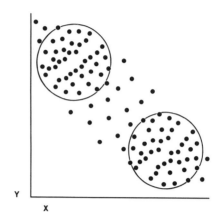

Figure 46. Groups in the data of figure 45 defined upon the basis of relative closeness of points.

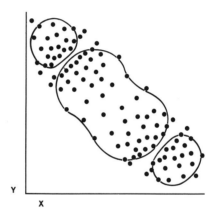

Figure 47. Groups in the data of figure 45 defined upon the degree of change of closeness of points.

evolution we may be thinking of similar elements that are defined through phylogeny. The second type of group, the segregate, may be equally "real" for other purposes, that is, within evolution certain forms of somewhat similar appearance are nevertheless segregated, perhaps in relation to phylogeny. In terms of data spaces it is clear that we may have dense areas close to other dense areas yet with narrow chasms separating them. The problem is essentially that the aggregate can be easily located by analytical methods (e.g., multivariate statistics) but the recognition of the segregate requires topological concepts and procedures. It is in this way that a new look can be taken at some biological data sets.

The words *aggregates* and *segregates* are not being put forward here as a terminology: first, at this stage, such terminology is not necessary because few if any workers within primatology are utilizing such procedures; secondly, they may well be mistaken for the carefully but differently defined mathematical concepts of "aggregate" and "segregate."

Some Properties of Data Spaces

Further clarification of this discussion is possible. For instance, a naïve consideration assumes that sets of biological data "exist" in isotropic spaces. By this it is meant that the data spaces possess the same properties everywhere. A simple analogy can express what this means in terms of a single original variable. Thus data may be plotted along a linear axis upon a rubber sheet (figure 48). If that rubber sheet is of uniform thickness, then stretching it produces changes between small values of x that are similar to those between large values. Stretching it in the x direction alone maximizes changes in x. Stretches in other directions have lesser effects on x but they are nevertheless uniform; nowhere in the rubber sheet do the effects of stretches differ from those in other places.

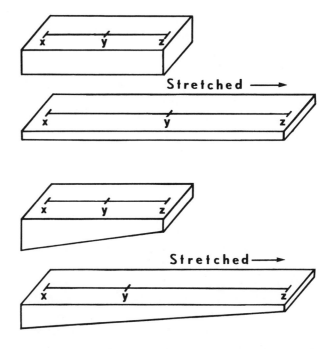

Figure 48. The rubber sheet analogy.

The analogy that is being suggested here is a "one-to-many" analogy. Thus data from a stretched rubber sheet as compared with that from an unstretched sheet may be thought of as being from a second specimen as compared to the first. This analogy suggests that the second set of data gives similar information as the first because the relative proportions within the sheets do not change with stretching. But the analogy may also be used to suggest another kind of comparison. If we assume that sets of data are taken on two groups of animals, the data set with its variances and covariances determined for the first group may be analogized to data taken on the unstretched rubber sheet. Likewise, the data on the second group of animals (which may also be represented in terms of variances and covariances) are those obtained from measurement of the rubber sheet in a stretched state. Again, the analogy is suggesting that the relative pattern of variances and covariances within and between the groups will be similar. And there are other more complicated analogies that we may derive in relation to what may be called "isotropic data."

Many former examples of data analysis do indeed assume such an isotropic nature of data. For instance, the procedure of adding variables to one another, done in the earliest days of numerical taxonomy, is treating data as though they are isotropic. Analysis of data (or shape) by eye often falls into this category because the eye has difficulty in dealing with correlation between characters, one of the ways in which data spaces may be anisotropic.

In fact, it is well recognized that most biological data sets actually exist within anisotropic spaces. In terms of our analogy above, this is the equivalent of saying that for the original variable x drawn upon rubber, the sheet is not of the same thickness everywhere. For instance, such a sheet may be thickest near the origin and may gradually thin along the direction of the x axis (figure 48). If we stretch this sheet, those distances along the x axis which lie near to the origin separate only a little; those which already lie a long way from the origin separate very much more. In this case the data may be thought of as lying in an anisotropic data space.

Again, we can look at this in terms of our "one-to-many" analogy. Thus differences between the data obtained from the tapered rubber sheet before and after stretching may be likened to differences between one specimen and the next within what is considered a single group of specimens. Here it will be clear that the relationships between the measurements within the first animal will not be the same as those within the second. This type of anisotropicity is, of course, remarkably common; it is what is seen when specimens actually belong to the same group but are different in overall size; it is an expression of allometric relationship. As before, we may also suggest a second analogy relating to the comparison of data obtained from two different groups of specimens. Again, the data from the first group is equivalent to measurement upon the unstretched tapered sheet; and that from the second, equivalent to measurement upon the stretched tapered sheet. In this case what is obtained are variances and covariances that differ markedly between our two groups of specimens. There are even more complicated situations to which our tapered rubber sheet analogy may be applied; the essence, however, is to derive a simple picture of "anisotropic data."

These concepts are well recognized, and with more sophisticated methods of analysis it is still possible to understand them. It is at this point, however, that an interesting dichotomy of investigative methods seems to present itself.

Thus the usual way of dealing with the anisotropic situation is to perform certain maneuvers with the original variables so as to transform them into new variates lying in an isotropic space. For instance, in the single variable example quoted above, depending upon the nature of the taper, a transformation may be applied.

Quite frequently a logarithmic transformation is appropriate. In other words, the data on a tapering piece of rubber is replaced by data on a piece of rubber that is even throughout.

In terms of a multivariate situation, pure noncorrelation (which must rarely exist in biology), is one expression of an isotropic data space; varying correlation of some degree or other is the equivalent of an anisotropic data space, the degree of nonhomogeneity being related to the nature of the correlation of the characters. In this case, then, the investigation of the data might involve a technique such as canonical analysis; here the information contained in the original variously correlated variables is transformed into information contained within new canonical variates which are not correlated at all with one another. In other words, the anisotropic data are analyzed by being transformed into isotropic data, so as to yield results more easily.

As we have already discussed, this procedure is valid only if the data are normally distributed and if a common variance-covariance structure can be assumed. In terms of our rubber sheet analogy, this means that the technique of changing from a specially tapered rubber sheet to an even rubber sheet can be performed *only if the initial degree of taper* (initial degree of anisotropy) *is regular throughout the data sheet.* This was also essential in our simpler example, in which the nature of that particular taper was required to be approximately regular (e.g., logarithmic). This regularity of anisotropicity may well be the case, and under these circumstances such transforming techniques are powerful indeed.

However, it is conceivable that "real biological data" may be both anisotropic and irregular. In terms of our analogy this means that the rubber sheet is of varying thickness throughout, and that stretching it produces within it a series of changes which are predictable only with detailed knowledge of the rubber sheet. If this situation is thought to exist, then it seems that the logical way to analyze the system is to determine the nature of the irregularity (anisotropy) rather than to convert the irregular rubber sheet into a regular one. In other words, we wish to know the precise nature of the irregularities of the original sheet.

And if we consider the situation in figures 45, 46, and 47, we can see that it may not be at all easy to determine whether or not an anisotropic situation is regular or irregular. In each one-dimensional axis (figure 45) the spreads of the x's and y's seem to approximate to normal distributions; yet the total two-dimensional view shows that the situation is not so simple. Moreover, such an overview reveals the two conflicting ways of looking at the data mentioned above. For instance, if we consider the idea of locality or distance as being the necessary view of the data, then it would seem as though the groups that exist are the aggregates (figure 46); they certainly contain the centers of density; they are approximately elliptical, and with appropriate transformation these groups might be analyzed by techniques such as generalized distance or canonical analysis. But if we consider, as the important parameter of distinction, the degree of change as one passes from one data point (specimen) to another, then the data take on an entirely different cast; they now seem to consist of the three segregates (figure 47). Here the distinction between the groups is the fact that though each is similarly dense at those points where they are close, and though density declines slowly as one passes in a direction away from interfaces with other groups, nevertheless the groups are sharply separated by regions in which there is a sudden change of density.

The first way of looking at the data provides two relatively high peaks separated by a somewhat lower plateau; the second way of envisioning the information demonstrates three groups that are sharply separated by rift valleys at those very points where the peaks are highest.

Therefore we are describing here a data space which is markedly and irregularly anisotropic; the example clearly shows that transforming it to an equivalent isotropic data space not only may hide one important aspect of the data (the presence of the rift valleys) but may even prove misleading by masking the second set of groups through releasing the first. In fact, we would like to have both ways of looking at the situation: the knowledge of the aggregated groups may be every bit as important as that of the segregated groups. But the techniques so far discussed are only able to view the former. Thus we have searched for methods which do not transform the data but, rather, which attempt to investigate the data space as it is, with the purpose of defining the various and changing natures of its anisotropicity (i.e., in terms of our original analogy, for discovering the irregularly uneven nature of the rubber sheet).

Group Finding Procedures

The attempt to solve problems when data spaces are irregularly anisotropic is thus, at least in part, one of describing aggregates and distinguishing segregates. Accordingly, it is to topology that we must turn. The techniques of this branch of mathematics are able to investigate the irregularities of the rubber sheet as it stands. The particular method (evolved by Peter Neely of the University of Kansas) is called Neighborhood Limited Classification.

As with multivariate statistics, the reader must approach appropriate texts (general: Estabrook, 1966; Abbott, 1969; specific: Neely, 1972; Oxnard and Neely, 1969; Oxnard, 1972a) for precise mathematical exposition. However, it is essential for the biologist to have a geometric notion of how the method works; this is visualized in the next series of diagrams.

Figure 49 shows a small number of points (specimens) lying in a plane, that is, characterized by two measurements taken on each. Although all the points may be defined as a single primary group, the eye readily discerns that three secondary groups are nevertheless present. (If the dimensionality of the data space is greater than three, then the groups cannot be distinguished by this technique). In figure 50 the neighbors among these points have been demonstrated by their "relative closeness" to one another (a rule, based, in this case, on simple Euclidean distances). Again, within the whole group of data points the same three subgroups emerge when we allow the criterion of "groupiness" to be relatively short neighbor connections.

Group Finding Procedures in Morphology

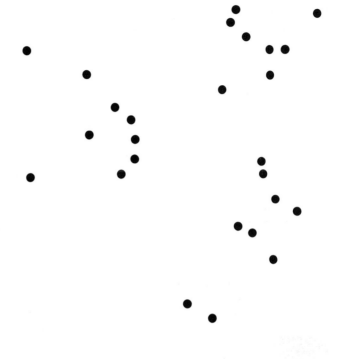

Figure 49. A set of data in a plane where groupings can be readily obtained by eye.

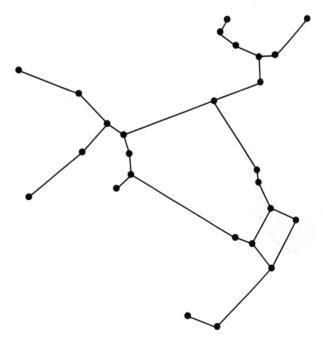

Figure 50. The same set of data with groups defined by relative closeness.

A more complex method of characterizing the data is the use of the barycentric decomposition. In this technique each point is enclosed within the convex hull of its neighbors; any point that would convert a convexity into a concavity is not considered as a neighbor; this is demonstrated in figure 51. As with our previous example, within the whole group of points the same three subgroups emerge, defined in this case by a considerably greater number of short and long neighbor links.

An even more detailed method is shown in figure 52, where we consider as neighbors any two points that are connected by an edge in a simplicial decomposition. With this technique one problem is to decide on alternative decompositions; for instance, a simple rule – which might be computationally impracticable – is to minimize the total length of the edges. Yet again the data are divided into the same three groups based, in this case, upon an even greater number of short and long neighbor links between points.

In all four of these examples groups are defined by the duality of looking for centers of high density (short neighbor links) and lines of low density between them (long neighbor links). In all four analyses, it is impossible to discern groups within the data if the dimensionality of the space is greater than three. However, in the last three cases, numerical data defining the actual lengths of the neighbor links are readily produced, and the techniques can therefore "view" the groups irrespective of dimensionality.

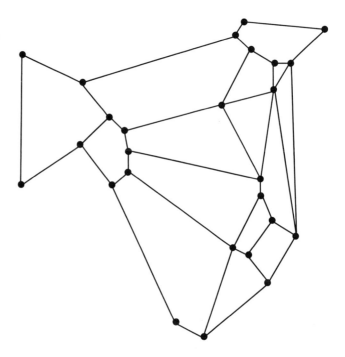

Figure 51. The same set of data with groups defined by a barycentric decomposition.

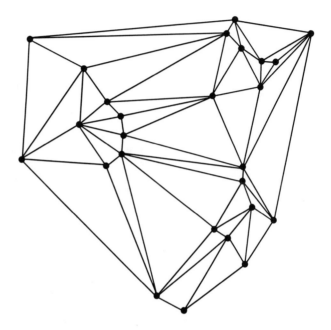

Figure 52. The same set of data with groups defined by a simplicial decomposition.

Figure 53 gives the sets of neighbors and their formation into groups as performed by the present state of neighborhood limited classification. The groups are directly evident in the trunk diagram and, as this does not depend upon dimensionality, they would be equally obvious in a multi-dimensional case. The comparison of these different techniques in this simple example shows how similar are the groups derived.

In order to see how the groups in figure 53 relate to the specimens in figure 49, it is easiest to consider an even simpler grouping of only six items (a, b, c, d, e, f). We can suppose that the group a, b, is formed with an association of 0.9 an arbitrary number reflecting distance (say). Next the group c, d, e, is formed at 0.7. These two groups are then joined to form a supergroup at 0.4 (a, b, c, d, e) and finally point f is added at 0.3. These results are indicated in the four steps of the next diagram (figure 54). A regular tree with a slender trunk is shown in the first quadrant of the diagram; a tree with a thick trunk is shown at the second; a trunk without branches is shown in the third; the computer trunk diagram is shown in the last quadrant of the diagram and is essentially the same except for a half-space shift to the right of the labels at the top. Thus an asterisk under a point indicates that it (or its group) is joined to the point (or group) at its right.

Other features of the method are displayed in figure 55. In this case, from a hypothetical set of data in a plane, we can see the various concepts that are associated with the method. Thus at a given level of association the method will define two groups A and B consisting of a and b specimens that are being joined by the interface. In each of the groups a boundary of points is isolated, as are the equivalent points of the interior; both are indicated in the alpha-numeric part of the

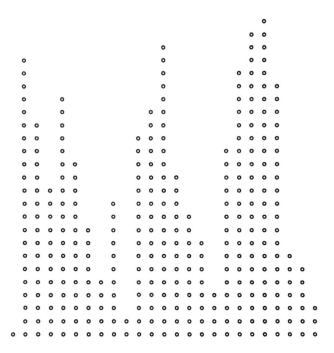

Figure 53. The same set of data with groups defined by neighborhood limited classification. In all of these figures (49 through 53) the same groups are derived. Neighborhood limited classification, however, provides a nondimensional view.

output. Every pair of links between neighbors in the interface is given. Outlying points which will be the next set of neighbors to the newly formed supergroup *A-B* are listed if they exist. This information is repeated in numerical form at every step of the grouping process. Thus the computer-drawn groups shown in the trunk diagram are merely visual agents for identifying the more detailed information.

Let us now consider the results of the application of this technique to a set of data which is more complex (because it contains larger groups that are of varying sizes, shapes, and densities) but which is still simple in the sense that the data lie in a plane and can therefore be assessed by eye. From the basic data shown in figure 56, it is apparent to the eye that there are three round groups of differing densities and one rather diffuse sausage-shaped group across the center. The next diagram (figure 57) shows the neighborhood limited classification of this data. The three relatively round groups are very evident, as is their relative tightness (one is big and tight; one small and very tight; one is fairly big but relatively loose). A final group is loose but is nevertheless evident as a group entity well separated from the others; the fact that it sprawls across the page in a manner that is anything but bivariate normal has not hidden the unitary nature of those specimens.

As yet another test, a set of generated random data is examined (figure 58) and the next diagram shows how neighborhood limited classification deals with this particular circumstance (figure 59). Groups are evident within the data; some are larger and others smaller; the random nature of the overall arrangement seems obvious.

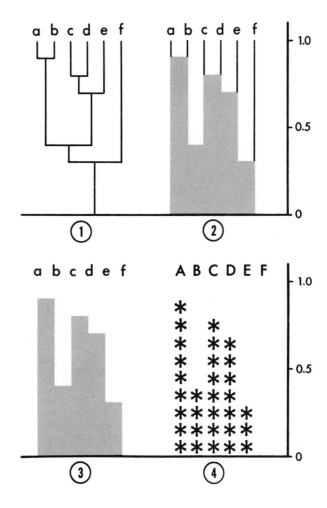

Figure 54. Tree representation of grouping process for simple data. 1 = regular tree with slender trunk; 2 = regular tree with thick trunk; 3 = trunk only; 4 = trunk printed by computer.

Application to the Shoulder in Primates: Evidence of Functional Information

Neighborhood limited classification has been applied to the data on the primate shoulder. One reason for such a study is to attempt to check, by means other than multivariate statistics, the functional interpretations derived from the canonical analysis of those data. Further motives, however, stem from our knowledge of the powers of the technique and from a desire to see how it performs with real biological information. Apart from the sets of simple artificial data, of generated random data, and of data consisting of a number of complex and noncircular groups as just described, the only data on which the technique has previously been tested are two sets obtained from the literature. Although "real" (one was obtained from observations on absolute magnitude and surface temperature of different stars, ostensibly an attempt to differentiate between white dwarfs and red giants; the other was of sepal and petal width of three species of iris examined by

Figure 55. A diagrammatic representation of other features of neighborhood limited classification.

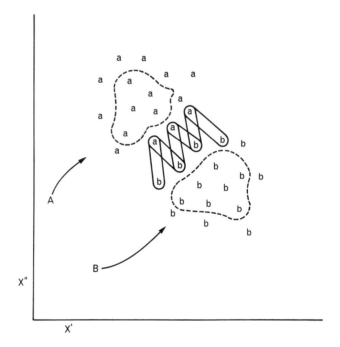

Figure 56. A set of hypothetical data in a plane demonstrating complex groups.

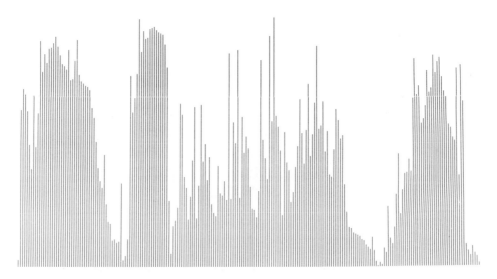

Figure 57. The groups derived from the data of figure 56 by neighborhood limited classification.

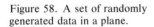

Figure 58. A set of randomly generated data in a plane.

Fisher—see chapter 1), these two were not genuine efforts to solve biological problems; rather, they were used as convenient data sets for establishing some of the properties of the technique (Neely, 1972).

When applied to the data on the shoulder girdle, the method produces overall conclusions (figure 60) similar to those from the canonical analysis of those data

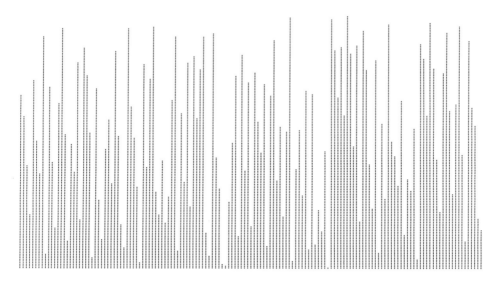

Figure 59. The groups derived from the neighborhood limited classification of the data in figure 58.

(chapter 3). Neighborhood limited classification suggests that the various specimens of the primates can be grouped into divisions that associate well with views about the function of the shoulder. It is able to differentiate between those monkeys that live primarily on the ground, where the shoulder is more frequently used in cranio-caudal movement, and those that are more highly arboreal, where there is a greater degree of movement of the shoulder in all three dimensions. These results are clearly evident from the diagram of the computer output. Group A comprises those monkeys that are essentially terrestrial (baboons, mandrills, and patas monkeys). Group B consists of those that are semi-arboreal or more fully arboreal but which do not display excessively acrobatic use of the shoulder in climbing and foraging in the trees (for example squirrel monkeys from the New World and macaques and mangabeys from the Old).

The technique is also clearly able to differentiate the less acrobatic arboreal monkeys from those that are more highly acrobatic in their use of the shoulder in climbing and foraging in the trees. Differentiated from group B is group C, which includes monkeys that are generally more acrobatic in their use of the shoulder in the sense that they sometimes (or, for some species, often) pull themselves up or even hang by the arms either during locomotion or while foraging, especially in positions where the shoulder mechanism is involved in abduction while in the raised position (as in woolly and proboscis monkeys). This differs, for example, from hanging positions in guenons (of group B), which suspend themselves more rarely and, when they do so, hold the arm in a flexed rather than abducted position.

Furthermore, the method differentiates a group D, which comprises all the apes (together with the spider monkeys and woolly spider monkeys of the New World). These species, however arboreal (e.g., gibbons) or terrestrial (e.g., gorillas) they

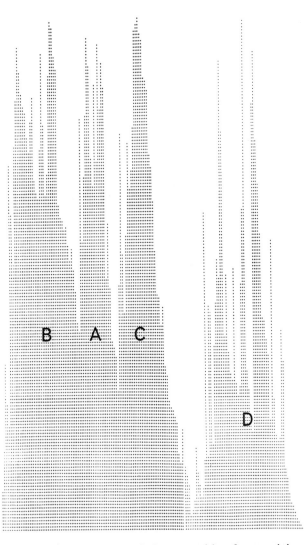

Figure 60. Computer display of the groupings within the locomotor (nine-dimensional) data from the shoulder girdle of the nonhuman Anthropoidea produced by neighborhood limited classification. Group A includes specimens of genera *Papio, Mandrillus,* and *Erythrocebus*. Group B includes specimens of *Aotus, Callicebus, Cacajao, Pithecia, Cebus, Saimiri, Callimico, Callithrix,* and *Leontocebus* from the New World, and *Macaca, Cercocebus,* and *Cercopithecus* from the Old World. Group C includes specimens from *Lagothrix, Alouatta, Presbytis, Pygathrix, Rhinopithecus, Nasalis,* and *Colobus*. Group D includes specimens from *Ateles, Brachyteles, Hylobates, Pongo, Pan,* and *Gorilla*. Exceptions to these groupings are discussed in the text.

may be in the totality of their locomotion, are nevertheless capable of arm raising (especially in the abducted position) and suspension of the body by the arms, such as is rarely displayed by other primates.

Finally, though not illustrated in figure 60, the technique distinguishes man from all other primates. All this information replicates that which was achieved by canonical analysis.

It is of considerable interest to know about relationships among the various groups revealed by neighborhood limited classification. How real are they? A question like this can be readily answered by examination of the more detailed numerical aspects of the computer output. Group A (terrestrial forms) is a very distinct group. Its component specimens are tightly located; its interior is large in proportion to its boundary (i.e., it contains many more specimens); its interface with the next group (*B* — arboreal monkeys) is small, consisting of only two neighbor links which are related to a single specimen of *Cercocebus* in group *B*. There are no specimens which are misclassified; that is, no arboreal specimens appear in group *A*. And of the possible specimens that are outlying to group *A* (i.e., have not yet been joined to that group), only one exists — a specimen of *Papio*, a terrestrial form.

At the other end of this analysis we have group *D*. What is the reality of this group? It is immediately obvious from the diagram that it does not coalesce until a much later stage than the others; in addition it seems to be considerably flatter and to possess a number of subgroups within it. Detailed examination of the output suggests the following: First, it is true that this group comprises those animals capable of extreme acrobatic activities at the shoulder (great and lesser apes, spider and woolly spider monkeys). Its discreteness from the remaining monkeys is not in doubt; thus the interface with them is at a low level of association (five steps from the end of the whole analysis — very distant), and it consists of only a small number of links. These are with specimens of *Lagothrix, Rhinopithecus,* and *Alouatta*. They are those particular members of *C* which are far distant from the interface of group *C* with group *B*. No specimens belonging (biologically) to other groups have been "misclassified" into the interior of group *D*. Conversely, there are no specimens of group *D* that have been "misplaced" into other major groups. The only specimens of genera represented in group *D* that are not themselves in the group are those which are outliers to group *D;* these comprise ten specimens in all. The separation of group *D* from the rest is thus very real. At the same time, investigation within group *D* reveals that the small-bodied acrobatic arboreal forms (gibbons, siamangs, spider and woolly spider monkeys) are a confused subgroup with closest links to a second subgroup of large-bodied acrobatic apes (*Pongo*). The terrestrial gorilla and chimpanzee (clearly capable of advanced acrobatics at present, and perhaps even a good deal more so in times past) form their own joint subgroup with links to the group of orangutans.

The investigation of groups *B* and *C* reveals a somewhat different picture. Here the diagram of the overall results indicates that two clear groups are present. But when the nature of those specimens within "wrong" groups or those that are outlying are examined, it becomes clear that the separation is not quite so marked as before. The interface in this case is still rather small, comprising only six links. However, specimens from one genus have been divided and are approximately evenly split between the two groups. Interestingly enough, these are the specimens of the low-canopy langurs which are clearly equivocal on behavioral grounds (see Oxnard and Neely, 1969, for further discussion). Twelve specimens which we might expect to see in group *C* were excluded at this point. Further investigation of the grouping process places these as outliers to the "correct" group, again except for two specimens of langur which are allied to group *B*. Thus a survey suggests that these two groups are also "real" but that their links with one another are somewhat more complex and extensive than is the case for the other groups discussed. This accords reasonably well with the behavioral (locomotor) actualities as far as they are known (see Oxnard and Neely, 1969; Oxnard, 1969c).

Finally, it is necessary to examine the overall relationship of these various groups as they exist in multidimensional space. The computer diagram is essentially nondimensional; that is, though it demonstrates the presence of groups, it does not reveal the spatial relationships between them. This can be obtained from the numerical part of the output. Therefore, though we cannot say at present what the full dimensionalities of the situation are we can say that a two-dimensional simulation of them is as given in figure 61. That is, group A does not lie between groups B and C as appears from the computer output. Rather, there is an order relationship among the groups $A-B-C-D$; but we cannot tell how "straight" that relationship may be. The order relationship is evident, however, because investigation of the interiors of each group shows clearly that the different interfaces contain only specimens that are at opposite edges of the groups. This information is not clearly evident in the canonical analysis of these data, although it is true that the canonical analysis of additional variables (all seventeen locomotor variables plus residual ones), together with cluster analysis of squared generalized distances, confirms these findings.

Perception of Irregular Groups

The use of neighborhood limited classification on the data from the primate shoulder also justifies one of the specific properties of the method: that it is able to perceive groups that may be of shapes other than spherical (or hyperspherical). In the canonical analysis of those data, one feature of the results is the lack of separate identification distinguishing a number of prosimian forms which often move in such a way that the shoulder bears tensile forces (e.g., potto, angwantibo), from certain New and Old World monkeys (e.g., woolly and howler monkeys) which utilize the shoulder in an acrobatic manner such that it bears tensile forces more frequently than in other monkeys. Neighborhood limited classification demonstrates a clear separation between these particular prosimians and monkeys in a rather interesting manner. The computer diagram of these specimens is shown in figure 62. It indicates two major groups, A and B, each of which shows obvious subpeaks within itself: A_1, A_2, B_1, and B_2. (The small group C is irrelevant to this

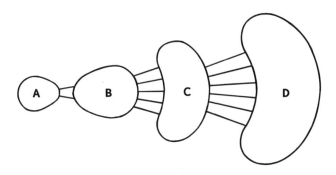

Figure 61. A two-dimensional representation of the links between the major groups A, B, C, and D produced by the numerical part of the analysis of figure 60.

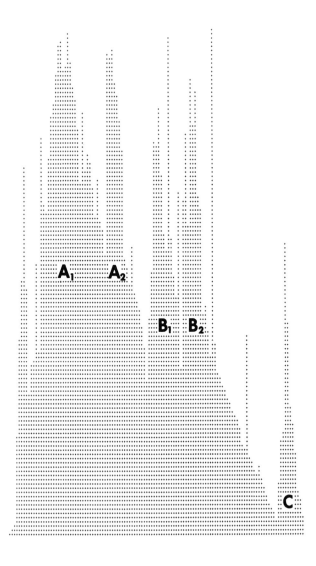

Figure 62. Computer display of the groupings within the locomotor (nine-dimensional) data from the shoulder girdle of the group of primates from those Anthropoidea and Prosimii which are acrobatic to an intermediate degree in their use of the shoulder in locomotion. These genera are not separated by the canonical analysis of these data; subgroupings are achieved by neighborhood limited classification.

discussion). At first sight the analysis is disappointing because each of the two major groups contains both prosimians and anthropoids. But closer study indicates that in each case the subscripted subgroups contain specimens essentially of only one taxonomic group: A_1 and B_1 contain primarily prosimian genera, while A_2 and B_2 comprise anthropoid genera. Thus, although the distribution of these forms does not allow a separation of all prosimians from all anthropoids, as an initial step, it does allow relatively easy identification of the specimens.

However, when we consider the interfaces between these groups, further interesting information appears (figure 63). The interface between A_1 and A_2 is very small, consisting of only one neighbor link; similarly, the interface between B_1 and B_2 consists of two neighbor links. This confirms the reality of the separations

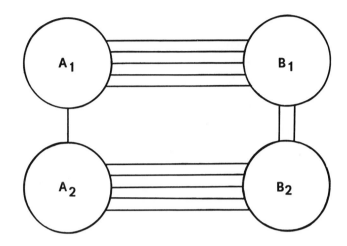

Figure 63. A two-dimensional representation of the links between major groups A_1, A_2, B_1, and B_2 produced by the numerical part of the analysis of figure 62.

between the anthropoid and prosimian subsets within A and B respectively. But the interface between the overall groups A and B is extensive, with ten pairs of links. And it so happens that half of these are between specimens in A_1 and B_1 (all prosimians) while the other half are links between A_2 and B_2 (all anthropoids). In other words, the nature of the groupings is such that within the multidimensional space the two taxonomic groups emerge as dumbbell-shaped. Each hyper-dumbbell comprises only specimens from its own taxonomic group. The lengths of the body connecting the heads of the dumbbells is greater than the distance between homologous heads on each dumbbell. However, the number of longer neighbor connections along the bodies of the dumbbells is almost an order of magnitude greater than the number of short neighbor connections from one dumbbell to the other. The groups that exist in this data are seen to include four aggregates (A_1, A_2, B_1, and B_2) arranged as two segregates A_1–B_1 and A_2–B_2. The total group A-B is essentially composed of two segregates, each lying rather close together but with "rifts" between them that are very steep. Neighborhood limited classification is able to reveal this anisotropic nature of the data space and thus separates the Prosimii and the Anthropoidea. Canonical analysis reads the situation as equivalent to many closely overlapping (hyper) spheres (genera) and is much less able to distinguish these specimens from the two suborders (though the canonical analysis of additional data achieves this; see chapter 3, and Ashton, Flinn, Oxnard, and Spence, 1971).

A second example of this nature can also be seen from the examination of the shoulder data. There seems to be even greater confusion between those Anthropoidea which are essentially quadrupedal whether in the trees or on the ground (e.g., squirrel monkeys and guenons) and those Prosimii which are likewise primarily quadrupeds (e.g., lemurs of various types). In this particular investigation tree shrews are also included as prosimian quadrupeds because they have been classified as primates from time to time; they are certainly quadrupedal in their general locomotion. Figure 64 shows the computer output of the neighborhood

Figure 64. Computer display of the groups within the data from the shoulder girdle of those primates that include the most quadrupedal members of both the Prosimii and the Anthropoidea. These genera are not separated by the canonical analysis of this data; subgroupings are achieved by neighborhood limited classification.

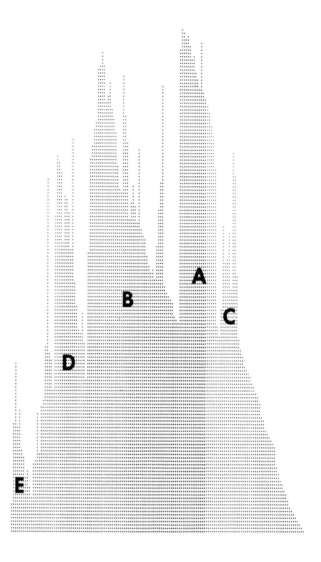

limited classification of these specimens. Here the immediate impression is that there is a single overall group of all quadrupeds; and it would seem that this makes considerable sense when we view the function of the shoulder in these animals. This group is a typical example of an aggregate. However, prosimians are different from anthropoids taxonomically and it would be of considerable interest if evidence of this were, in fact, hidden within the aspects of the shape of the shoulder as a series of segregates.

We have a special interest, therefore, in the subgroups A, B, C, D, and E, of the single major block. First, subgroup A is a clearly defined group of specimens that are very tightly associated with one another; they comprise every single specimen of tree shrew available, and include no other specimens. The second group *B* is somewhat more diffuse, containing specimens from both the Anthropoidea and the Prosimii. Groups *C*, *D*, and *E* contain only specimens of the Anthropoidea. This is a less satisfactory sort of separation than was achieved in the previous analysis. For in this case, while tree shrews and some monkeys can be distinguished, the remainder of the forms are apparently still confused.

Study of the interfaces between these groups confirms the very distinct nature of the tree shrews; there is only a single link between them and the rest. Interfaces between groups *C*, *D*, and *E* show that they are arranged in an interlocking fashion. *C* has three links with *D* and three links with *E*; yet *D* and *E* are close enough to one another to possess a single link between them. These particular anthropoids are therefore fairly well defined. Interest centers on the combined group *B* and the nature of its interior and of its interfaces with the other groups.

It has already been noted that group *B* is related to group *A*, the tree shrews. Study of the single link interface between them shows that it is between a tree shrew and a lemur. Group *B* is also related to the triad of groups *C*, *D*, and *E*, but of these the one with neighbor relationships is group *C*. There are 13 links between group *C* and group *B*; in almost every case these are links with the anthropoid members of group *B*. Within group *B* there is a locus of separation between the prosimians and the anthropoids in the sense that the prosimian members are rather tightly organized and eccentrically placed so that they lie at one edge of *B* (the edge with which the tree shrews have contact). The anthropoid specimens are less tightly organized and at the opposite edge of the group. Study of the outlying specimens confirms this picture. All those outlying specimens which are of the Anthropoidea (22 specimens) are outliers to *C*, *D*, and *E*, or, when outlying to *B*, are nearest the anthropoid members of that group; almost all those outliers (17) which are prosimians are related to the prosimian elements of group *B*. Hence it is possible to diagram these relationships within two-dimensional space as in figure 65. It is clear that there are examples not only of both aggregates (the whole group) and segregates (*A*, *C*, *D*, and *E*) but also of a group (*B*) that is intermediate between them. This more complex separation is not achieved by canonical analysis of the same data, although again its presence is confirmed by analysis of additional data (see chapter 3).

Morphological Peaks and Troughs

The new technique provides information relating to the formation, from the shoulder data for individual specimens, of markedly obvious suprageneric groupings or peaks. Not only are these peaks related to data from specimens that are closer together at the peaks rather than between them, but the same also applies to the genera themselves. Genera are crowded together at the peaks, and those genera closer to the peaks have smaller spreads than do those which lie away from or between peaks.

The study of the function of the shoulder within locomotion does not seem immediately to fit this last result. For instance, although it is possible to suggest that the function of the shoulder in different primates can be grouped in terms of

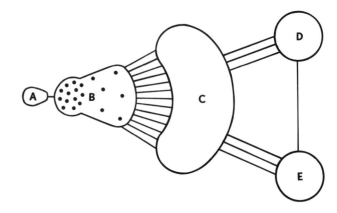

Figure 65. A two-dimensional representation of the links between the major groups A, B, C, D, and E, produced by the analysis of figure 64.

how the animals are thought to move, it must also be recognized that any such groupings are liable to be arbitrary.

First, this arbitrary nature is a consequence of ignorance about the details of shoulder function in locomotion in a large number of the rarer primate genera. For instance, rather little is known about the locomotion (and more widely, the behavior) in the field of genera such as *Cacajao, Pithecia,* and *Chiropotes* in the New World, and of genera such as *Pygathrix* and *Simias* in the Old; among prosimians almost nothing is known of the habits of *Daubentonia;* and there are many other genera for which reasonable information is lacking.

Secondly, its arbitrary nature results from the attempt to suggest overall groupings for genera which are, after all, generic collections of species that may well move in different ways. For example, within the genus *Cercopithecus* there are known variants ranging from forms like *aethiops* which spend relatively longer periods of time on the ground, through forms like *talapoin* which frequent low bushes and undergrowth, through many of the regular cercopitheques which move freely at all levels from ground to canopy, and finally to species such as *diana* which spend most time in the high canopy (Tappen, 1960). Within the genus *Presbytis* it is well documented that there are some (e.g., *entellus*) that spend more time on the ground than do others (e.g., *obscurus*). Similar findings are relevant for mangabeys and macaques, and in the New World for howler monkeys; presumably as we come to know more about different species this intrageneric variation will be even more obvious and widespread.

Thirdly, even when there is considerable information about the range of locomotor behavior of species within genera and within given species, the very plasticity of the primates almost denies attempts at classification. Thus, almost all primates are capable of almost all ranges of locomotion. Insofar as overall locomotion is concerned exceptions include the ricochetal brachiation of gibbons and siamangs (and even this is approached by spider monkeys). In terms of the functions of the shoulder within locomotion, a greater number of differences among primates are apparent. For although all primates can and do hang by their forelimbs, only certain primates (for example, the gorilla and howler monkey) can do this with the shoulder fully raised in abduction; others do it with the shoulder in a flexed (protracted) position but with little abduction.

It seems, therefore, that the behavioral data suggest that there is a spectrum of behavior; yet the morphological data demonstrate that this undoubted range has been achieved (in evolution) in terms of discontinuities between one morphology and another. Does this suggest that in some way or another these morphological peaks may be related to such concepts as adaptive peaks? As can be seen from the figures, this morphological clustering is most marked; it completely overshadows any generic or other taxonomic grouping that we might expect to discern in the data.[1]

In retrospect it can be seen that canonical analysis vaguely foreshadows this result. But in that case, the fact that generic groupings are already given considerably obscures the picture. Although there is evidence of clustering of genera from such a study, there can be no evidence of clustering of specimens or of differing generic ranges correlated with positions of clusters.

The Combined Use of Multivariate and Cluster Analyses

The controversy that accompanied the earlier growth of the use of clustering methods in taxonomy has largely died down. These methods, included under the rubric of "numerical taxonomy," are, of course, of far wider application than in taxonomy itself. Many studies have now been carried out on a diversity of data. The literature and the range of methods have grown so extensively that few if any researchers can hope to keep up with them. Numbers of workers are now producing comparative surveys of methods attempting to discuss their biological rationale and value to a variety of problems. Finally, and most importantly of all, mathematicians themselves have begun to look seriously at these techniques as a worthwhile field of investigation; as a result, the near future will see considerable improvements, redundancies made obvious, and statistical properties of methods carefully defined (e.g., see Jardine and Sibson, 1971).

---

1. It is interesting that the preliminary examination of the canonical analysis of the data on the innominate bone (chapter 3) seems to suggest a similar result. In this case canonical analysis produces fairly obvious groupings that make considerable functional sense when viewed in the light of pelvic function in locomotion. At the same time, however, the groupings have created a number of somewhat curious bedfellows (e.g., gibbons and baboons); one explanation of this could reside in a similar set of preferred morphological peaks to which many different genera might conform even when functionally the animals might be doing somewhat different things. In other words, although natural selection is undoubtedly selecting those morphologies most adaptive to particular locomotor patterns, it may be doing so within certain constraints as to the morphologies that are possible. Thus, while it is likely that selective factors for mechanical efficiency may be partial causative agents, it is possible that the "genetic models" of the primate shoulder and pelvis respectively may have been sufficiently "fixed" at early evolutionary stages as to limit grossly the independent ways in which they may vary.

Such a hypothesis, although somewhat unusual, has been suggested by a number of studies: for instance, in mammals, on the shoulder (Oxnard, 1968) and cranial base (DuBrul and Laskin, 1961), and, in fishes, on certain jaw mechanisms (Liem, 1970); the fact that the current study on the pelvis has also shown similar constraints in shape in primates, must be reckoned as an additional support of these general speculations.

It is immediately obvious that these techniques can be applied, indeed should be applied, in the analysis of morphometric data resulting from investigations of adaptation in primates. The use of one such method, neighborhood limited classification, as a primary factor in research strategy has just been described (Oxnard, 1969c). But clustering methods can also be used as an aid in the understanding of data which has already been synthesized by multivariate statistics. Though such methods cannot actually show the multidimensional structures that exist, they can enumerate those structures within the data. This means that whatever clustering method is used, it immediately takes on the limitations of multivariate methods; however, given the well-understood nature of these methods, and their apparent robusticity in practical use, this is not necessarily an important limitation. It further follows, however, that if we wish to superimpose a clustering method upon multivariate statistics, we do not need to use a method like neighborhood limited classification, which is designed to circumvent the concepts of multivariate statistics themselves and which, it is freely admitted, is itself relatively untested. That particular method is rather specific, it depends upon a special mathematical background, and in a number of respects differs from the sorts of techniques that generally go under the heading of clustering analyses. Accordingly, we can use the more orthodox methods.

Here primate morphologists have entered the field rather late and at a stage when improvements of some of the techniques and deletion of others are imminent. We ourselves have therefore utilized only the most simple and best understood analyses (Ashton, Flinn, Oxnard, and Spence, 1971). Thus following the Rothamstead group (Gower, 1967; Gower and Ross, 1969; Ross, 1969) we have used the minimum spanning tree and single linkage cluster analysis. These have been applied as adjuncts to generalized distance statistics for displaying the results that may be hidden in a large matrix of such distances. Tentative mention of this has already been made in chapter 3. We are interested in viewing the clustering of the means of genera, assuming that intrageneric variation has been taken care of by the multivariate method.

Biological reviewers of clustering methods have pointed out, and rightly so, that a principal defect of single linkage cluster analysis is that it tends to form chains that are long and thin (Sokal and Sneath, 1963; Moss, 1968; Boyce, 1969). This criticism has also been applied by some of the mathematical reviewers who have evaluated the technique. Thus Wishart (1969) agrees with Lance and Williams (1966) that the chaining effect renders virtually obsolete such nearest-neighbor techniques as single linkage cluster analysis; he has pointed out, however, that this chaining effect does not obtrude if the data do possess very distinct clusters of any shape. Gower and Ross (1969) have further investigated the properties of a number of these clustering methods. They feel that single linkage cluster analysis may be of considerable value; for instance, knowledge about long clusters, if actually present in the data, may be informative; unlike most other methods, single linkage cluster analysis produces the same result when sorting upwards (agglomerating small clusters into large ones by increasing the clustering distance), as when sorting downwards (dividing large clusters into small ones by decreasing the sorting distance); and the single linkage cluster analysis can be readily obtained from the minimum spanning tree. This latter has been shown by Gower (1967) to be the common base of a number of clustering methods that on the surface are apparently rather different. Certainly a number of biologists using the techniques have shown that the main clusters given by a fairly large number of methods are generally similar. Variations are found to occur only in the more detailed aspects of group formation (Boyce, 1969; Hall, 1969a, b; Moss, 1968).

Pending further statistical investigations of these various methods, we have used the minimum spanning tree as a basic tool for examining generalized distance statistics, with the information contained within the single linkage cluster analysis as ancillary. The primary source of information remains in the full matrix of the generalized distances themselves. Our knowledge of the canonical variates is also at hand, so it is most unlikely that the cluster analysis we have used will lead us very far astray.

The minimum spanning tree is a method of structure determination within a set of points. It is that network connecting the points that is of minimum total length. Thus every point is joined by some path to every other point and no closed loops occur. Exactly $n - 1$ links are required to connect n points. Several computational techniques are available. Single linkage cluster analysis attempts to group the points within a multidimensional space into usually disjoint sets which, it is hoped, will correspond to marked features of the sample. The grouped sets of points may themselves be formed into larger groups so that all the points are eventually hierarchically classified. This hierarchical classification can be represented diagrammatically as a dendrogram.

The disadvantage of the minimum spanning tree is that it provides no information about how the various branches of the tree should lie relative to one another. This can be overcome for a small tree by drawing it on the vector diagram provided by a multivariate statistical technique. For certain trees the minimum pathway may be so close to certain other pathways that minor perturbations in data may radically alter the tree. In such cases relevant information may well be displayed by showing other near-minimum links. This may be done visually, or, for larger trees, a list of the few nearest neighbors of each object may be useful in planning how the branches should be drawn.

The principal value of the minimum spanning tree is that all the information required for the single linkage cluster analysis is contained within it. The various algorithms for finding the minimum spanning tree are efficient even when there are very many data points as in most true biological situations: for instance, in our efforts to investigate the structure of the shoulder, there are in one study more than one thousand specimens (Oxnard, 1968a), though it is true we have rarely measured more than seventeen variables on each specimen; in contrast, in the studies of Boyce (1969), where several different techniques are used in order to investigate the techniques themselves rather than a real biological problem, a great many more measurements were taken but on only 20 individuals. Simple methods like single linkage cluster analysis may therefore be of value in the practical situation when other methods are not.

The studies on the shoulder are able to supply examples of the use of clustering techniques in conjunction with multivariate statistics – in particular, of the way in which unsuspected information may be recovered from the data.

First, one may contrast the results of cluster analysis of generalized distances with those of canonical analysis by comparing the dendrogram resulting from the single linkage cluster analysis with the groups that seem to be contained within the plot of the major canonical axes. This has been done for the shoulder study and is shown in the figures 66 and 67. The principal finding is that the cluster analysis confirms the major groupings that appear to exist in the canonical analysis; this is notwithstanding the fact that considerable data are still contained in the remaining fifteen canonical variates (especially, of course, the 3rd and 4th) of that study. However, the technique does suggest additional information in that, though the various quadrupedal monkeys from the New and Old Worlds are not separable in the bivariate plot of canonical variates one and two (or indeed even when canonical axes three and four – not illustrated – are also considered), the cluster analysis shows clearly that the generalized distances do indeed separate New World forms from those of the Old World. A similar finding is also apparent when one looks at the cluster separations of the more acrobatic monkeys; canonical analysis does not clearly separate New and Old World forms.

This does not mean that canonical analysis is less efficient in some way or other than generalized distance analysis; as we have seen this cannot be so because the basis of the analysis is the same. It merely means that because particular views of the multidimensional model are forced upon the data in the choice of the canonical variates, smaller separations that do not happen to coincide with those views may well be spread piecemeal amongst a number of canonical variates, and hence hidden. Clustering the generalized distance values enables us to retrieve the hidden information. These particular results are of value because they may enable an unknown (fossil) form to be placed with even more accuracy. They certainly reveal a component of the shoulder data which coincides with the taxonomy of the extant forms.

A second way of comparing canonical analysis with the cluster analytic results of generalized distance analysis is to plot the clusters directly upon the background of orientation supplied by the bivariate plot of the first two canonical axes. In this case it is the minimum spanning tree which can be plotted. And though the minimum spanning tree is a unique set of connections within the groups, it is not necessarily biologically of most significance. The minimum spanning tree depends upon the actual minimum distances, but in any given case it may well be that a number of distances are all very similar; under such circumstances the choice of the actual minimum may merely relate to "noise" within the data. Hence, though the minimum spanning tree can be plotted as shown in figure 67, it is also possible to indicate the situations where a number of links are almost the same length. This may be of especial importance in examining relationships.

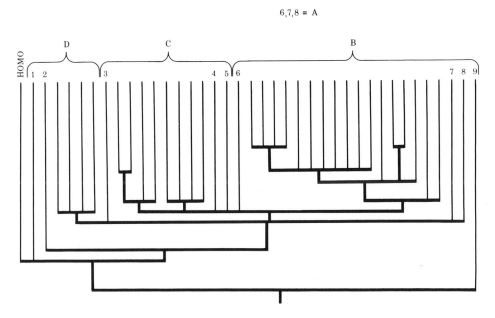

Figure 66. The dendrogram resulting from single linkage cluster analysis of all seventeen dimensions taken on the shoulder girdle of primates. The groups clustered in this analysis are the different primate genera (not the individual specimens, as is the case for neighborhood limited classification). The basis of clustering is the generalized distances of the means of genera, one from another.

Genera 6, 7, and 8 are the three terrestrial monkeys that are clustered together by the neighborhood limited classification of the smaller data set (these genera are the group *A* of figure 60).

Group *B* includes all those genera of both the Anthropoidea and Prosimii that may be grouped under the rubric arboreal quadrupeds (group B of figure 60).

Group *C* includes all those genera of both the Anthropoidea and Prosimii that are acrobatic to an intermediate degree in their use of the shoulder in locomotion (semibrachiators and hangers of a previous terminology). *Alouatta* (3) is especially noted (c.f. group C of figure 60).

Group *D* includes those forms, greater and lesser apes, spider and wooly spider monkeys, that display the most acrobatic capabilities of the shoulder among the primates. *Hylobates* (1) and *Pongo* (2) are especially noted (c.f. group D of figure 60).

Daubentonia (9) is the most outlying genus.

Examples of these phenomena are available in the analysis of the primate shoulder. Grouping the genera on the basis of canonical analysis alone results in the formation of the three obvious groups shown in chapter 3 and in figure 67. Each of these groups contains such a mixture of genera that, though the contained information makes some sense when viewed in the light of the function of the shoulder, it is apparently disappointing when viewed with knowledge of the taxonomy of the group because the criteria for taxa are more complex than just one locomotor function. That some information about taxonomy is present in the data is already known from the neighborhood limited classification of a similar (albeit smaller) data set. The clustering methods outlined in this section show how taxonomic information may sometimes be obtained from the multivariate statistical approach.

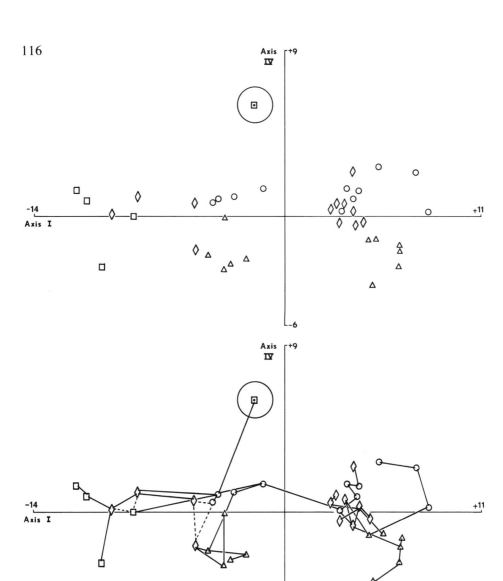

Figure 67. The combination of canonical analysis and the minimum spanning tree. The minimum spanning tree is based on the bivariate plot of canonical axes 1 and 4 for the seventeen–dimensional data on the primate shoulder. The circle around man (radius one standard deviation unit) gives a measure of the separation of the forms. Squares represent Hominoidea; diamonds represent New World monkeys; circles represent Old World monkeys; triangles represent Prosimii.

The top diagram shows the three major groups that are generally realized by the canonical analysis of the data: far right, the most quadrupedal forms; center, those that are intermediate; far left, the most acrobatic forms (in terms of the use of the shoulder). Man is uniquely different from all.

The lower figure is a similar plot but with the minimum spanning tree derived from the analysis of the generalized distances superimposed. This shows that additional information is included in the later canonical axes: for instance, within the general quadrupedal group (far right) Old World forms tend to be preferentially separated from New World forms, Prosimii tend to be separated from both of these, and the three genera of terrestrial monkeys form a separate tail at the most extreme right. Similarly, within the intermediate group, Prosimii and New and Old World monkeys form three subgroups. For further explanation see text.

Thus a dendrogram derived from single linkage cluster analysis of the seventeen-dimensional data on the primate shoulder reveals the picture shown in figure 66. Here the main groups (the quadrupeds, semibrachiators, and brachiators shown in earlier and cruder groupings, and revealed also by canonical analysis) are clearly present, but within each there is a series of subgroups that separate the different taxonomic groups of the primates from one another. However, the information conveyed in the dendrogram is not entirely clear, mainly because a dendrogram cannot convey neighbor relationships between forms. For instance, *Homo* and *Hylobates* appear from the diagram to be next to one another; this is an artefact of the way the dendrogram is prepared; in actuality *Hylobates* is nearest to group D (other apes, etc.), while *Homo* is uniquely placed so as to be approximately equidistant from several groups. Similarly, in figure 66, genus number 6 (*Erythrocebus*) appears to be widely separated from 7 and 8 (*Papio* and *Mandrillus*, terrestrial quadrupeds). This is a spurious impression. Numbers 6, 7, and 8 form a linear sequence at one extreme of the whole distribution of genera; of these, number 6 is slightly nearer to other quadrupedal forms and hence, sorts out in the dendrogram in a way that makes it appear totally different from 7 and 8. *Daubentonia*, number 9, is shown as the most widely divergent genus and this is directly contradictory (apparently) to the information given by the first three canonical axes; for further discussion of this point, see chapter 7. None of these more complex and important relationships is obvious from viewing the dendrogram.

The complexities are seen more succinctly if viewed utilizing the minimum spanning tree superimposed upon the picture displayed by canonical analysis (figure 67). The separations within the major groups shown by canonical analysis are obvious. But finer distinctions can be made within them, and the neighbor relationships of each genus can be clearly seen. Of course, in a diagram such as this the only "distances" that are "correct" are those represented as links. A number of other distances almost as "short" have also been figured as dotted lines. They indicate clearly where a minimum spanning tree is not particularly different from certain other sets of links. In most cases, however, the minimum spanning tree is by far the best. It is significant that in those cases where the minimum spanning tree is clearly unique, the groups which are formed make biological sense. Thus the group of terrestrial quadrupeds, not clearly separated from the other quadrupedal forms by canonical analysis, are nevertheless shown to form their own outlying "tail" to the general distribution of quadrupedal forms (figure 67). This confirms the information, obtained from the neighborhood limited classification, that these forms do indeed form a group within the general distribution of forms. This mode of viewing the information also clearly recognizes that (a) within the group of quadrupedal forms, the Prosimii can, on the whole, be differentiated from the Anthropoidea; (b) within the more acrobatic forms the Prosimii are separate from the Anthropoidea; and (c) within both of these groups, Old World and New World monkeys tend to be separated. Thus this analysis of the seventeen dimensions of the shoulder makes many taxonomic divisions which seem sensible and which could not be discerned from the canonical axes alone. It is of extreme interest that all of these are foreshadowed by the more powerful technique of neighborhood limited classification when this is applied to the smaller data set (nine locomotor dimensions, figures 62 through 65).

In some of those cases where the minimum spanning link is only slightly shorter than some other possible links, it does not always make the best possible biological sense. Figure 67 shows that although one member of the Hominoidea is preferentially linked to one of the less acrobatic New World monkeys by the minimum span, it is also very little further away from two other members of the most acrobatic group. In other words, this set of near minimum linkages emphasizes some of the boundary positions that we expect to find.

There are still other examples of the combined use of generalized distance analysis and canonical analysis through the medium of the minimum spanning tree and single linkage cluster analysis. Because they are related to the interpolation of unknown or fossil forms, they are dealt with in chapter 7.

A Note of Caution

The main theme of this chapter has been the use of a variety of group-finding techniques, none of which has been as well thought out, tested, and used as have the various methods of multivariate statistics. In particular, neighborhood limited classification has been limited to these particular investigations, and it is possible that the method may eventually be either much improved or discarded in favor of others.

It is not possible for me to supply theoretical discussion or statistical testing of techniques like neighborhood limited classification. This must await developments and criticisms by those qualified to provide them. But it is possible for me to apply those kinds of tests that depend upon the examination of the same set of data by means of several different methods. Such methods of testing do not supply new insights into the modus operandi of a method, but they do allow comparison in working situations with other perhaps better known methods.

Accordingly, a set of data on the primate shoulder which we believe we already understand fairly well has been studied in the following way. The data are those presented by the original nine dimensions taken on the shoulder girdles of those genera of the primates representing the Hominoidea, that is, *Homo, Gorilla, Pan, Pongo,* and *Hylobates*. The examination by canonical analysis of these five given groups yields the results shown in figure 68. The mean position for each genus is shown, and a 90 percent circle is inscribed around them. The results suggest that man is uniquely separate and that overlaps among the nonhuman genera are small. In particular, the genus *Hylobates* occupies a central position in the canonical space between *Pongo* and *Pan*.

When, however, we apply neighborhood limited classification to the same data, the groups that are discernible are those shown in figure 69. In this case also man is uniquely separate from the nonhuman Hominoidea. Most of the specimens of orangutans are also clearly defined. Two further groups, gorillas and chimpanzees, are evident, but here four specimens are misclassified. This degree of misclassifi-

Group Finding Procedures in Morphology

Figure 68. Canonical analysis of hominoid scapular data, axes one and two. Locomotor dimensions. Groups of genera are given; therefore *Pongo* is separate from *Hylobates*.

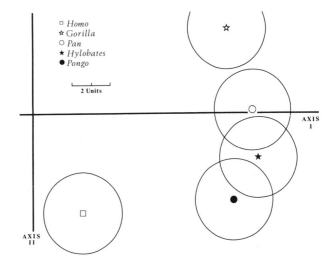

Figure 69. Neighborhood limited classification of the same data as figure 68. Groups are defined by the computer-drawn trunk diagram. Relationships between groups are represented in diagrammatic form below. Groups of genera are found; *Pongo* is inseparable from *Hylobates*.

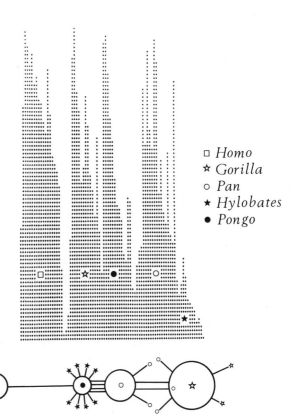

cation is entirely consonant with the overlap of limits in the previous example. The computer-drawn diagram suggests that a further small group exists, but inspection of the numerical part of the output shows that this is spurious; it includes specimens of *Pongo* and *Hylobates* that are outlying to the main group of *Pongo*. In other words, with this particular analysis it is not possible to separate *Pongo* and *Hylobates*.

Which study are we to trust? Clearly the answer to this question is important, because if we suppose that *Hylobates* were a fossil form, our assessment of the resemblance of its shoulder (size having been taken into account) to that of other forms would differ. Canonical analysis suggests that, in this respect, gibbons are phenetically intermediate between orangutans and chimpanzees; neighborhood limited classification implies that gibbons are phenetically closer to orangutans than chimpanzees.

If we enquire into possible functional meanings of these resemblances, other interpretations are possible. Speculation on the function of the shoulder through morphological resemblance on the basis of the canonical analysis suggests that gibbons are intermediate between highly arboreal orangutans and mostly terrestrial (though capable of acrobatics) chimpanzees. Neighborhood limited classification, on the other hand, suggests that gibbons are highly arboreal and acrobatic (as are orangutans) rather than relatively terrestrial like chimpanzees.

If we believe that the data provide taxonomic affinities, we will again have two possible speculations. We may either assume that neighborhood limited classification demonstrates gibbons to be taxonomically closer to orangutans, or we may feel that canonical analysis shows gibbons to be intermediate between orangutans and chimpanzees.

We have tried to test these various alternate explanations by applying further analyses that have become available. First, we have attempted to view the Hominoidea as seen through a discriminant function technique but where the actual specimens of each genus (rather than the means for each given group) are plotted in the eigenvector space. The results obtained from this procedure are shown in figure 70. The human eye readily places into the plot of the specimens the three straight lines shown, and these suggest that in reality only four groups exist: one, *Homo*, uniquely separate; two, *Gorilla*; three, *Pan*, having slight admixture with *Gorilla* across a narrow boundary; and four, *Pongo* and *Hylobates* combined. This last group is anything but round; those specimens that appear to interdigitate with *Pan* actually lie below that group (axis 3, not figured). This result is reminiscent of that provided by neighborhood limited classification, although it suffers from the deficit that the groups have been determined by eye.

Secondly, we have utilized the group-finding procedures of Rubin and Friedman (1967). These combine an elegant hill-climbing cluster procedure with data that have already been transformed (in a multivariate statistical sense) to the eigenvector space. The separations that are achievable are shown in figure 71 and suggest strongly that the combined group of *Pongo* and *Hylobates* is a reality. Gibbons and orangutans are sorted together, and as a group they are totally separate from chimpanzees. Man is unique in this analysis; the only overlap is

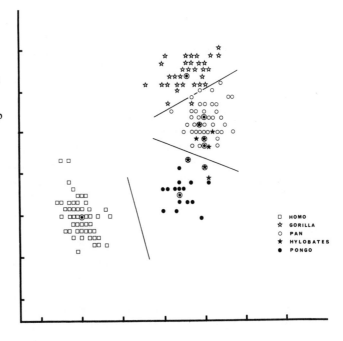

Figure 70. Discriminant function analysis for individual specimens based upon the same data as figure 68. The straight lines have been entered by eye between the groups. The apparent overlap of *Pan* on the one hand and the combined group of *Pongo* and *Hylobates* on the other is less real than apparent here, because a third dimension adds considerably to the separation. *Pongo* and *Hylobates* are not obviously separable.

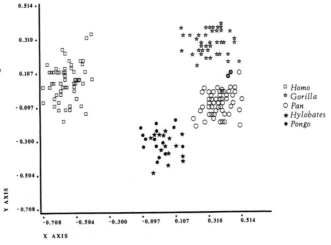

Figure 71. Cluster analysis in eigenvector space based upon the same data as in figure 68. Here the groups have been determined by the computer program. *Pongo* and *Hylobates* are not separated.

between gorillas and chimpanzees. This result is totally in accord with that presented by neighborhood limited classification (figure 69). It suggests that the apparently different result given by canonical analysis is because the groups are given rather than *found*.

This test is still only minor evidence; much remains to be done. Presumably the differences between the different results do not lie in the correctness or incorrectness of the methods; rather, they may reflect differing degrees of sensitivity in different situations.

6 Functional Significance of Bone Form: Experimental Stress Analysis

Introduction

In functional assessments of bone form and architecture, anatomical inference resulting from intelligent examination of behavioral and anatomical observations, is an essential tool. When modern methods of increasing the efficiency of the observations (e.g., by measurements and mathematical analysis) can be added to such compound observations, very powerful techniques are created. However, notwithstanding increased complexity and power, the deductions so formed remain correlative; their direct testing requires investigation of the function of the bone-joint-muscle complex and the biomechanics of movement. Such investigation is capable of confirming (which it usually does) or of denying specific anatomical inferences. But this is not the sole (or even possibly the most important) contribution of such methods; of further value in such studies is the new information and insight which cannot be obtained in any way from anatomical deductions. The epitome of such methods is the study of movement in vivo utilizing (a) telemetric devices that allow relatively unfettered primates to be studied; (b) photographic and television-like devices that allow simultaneous and detailed study of movement; and (c) stimulating devices that allow in vivo alteration of physiological parameters. With such instruments it is possible to monitor and experimentally alter, either separately or simultaneously, such physical and physiological properties of locomotor units as the amount of tension in tendons (e.g., using the "buckle" transducer: Salmons, 1969), strain in bones (using instruments such as the strain gauge rosette), and, perhaps most importantly, the electromyographic recordings of the electrical activity of muscles (Basmajian, 1972). Many of these techniques are challenges to the technology of our times. They can scarcely be used save by bio-engineering research teams. In many cases they are being provided only as a side result of medical and space instrumentation. Yet it must be acknowledged that such studies are being initiated in evolutionary investigations (e.g., Gaunt and Gans, 1969; Liem, 1970; Jenkins, 1970).

Other investigative modes are capable of supplying information bearing upon the functional significance of structure. Thus, at a theoretical level, simulation of bone-joint-muscle interactions may be carried out. Here computer models may give insight into structural parameters. For instance, simulation studies of the effects of changes in structural and physiological properties of muscles on muscular-induced movements (e.g., studies of degrees of pennateness of muscles, of relative positions of muscular attachments, of effects of series elastic elements in muscles) are revealing that the biomechanics of muscular action is considerably more complicated than is suggested by considerations of muscle power and lever arms alone (Stern, 1971b, and personal communication).

At a different, experimental, level it is also possible to investigate the mechanical efficiency of skeletal form by utilizing some of the analogical methods of experimental stress analysis. Some of these methods (strain-gauge circuits, brittle-lacquer and photostress coatings, Moiré fringe-strain analysis) are able to supply information about mechanical effects in actual objects. Their development rests upon appropriate engineering researches, and they are well described in a series of monographs on experimental stress analysis (e.g., Hetényi, 1950; Dove and Adams, 1964; Dally and Riley, 1965; Holister, 1967). The techniques have been used in a number of biomechanical studies: (a) strain-gauge techniques in a study of stresses within the femur (Singer, Milch, and Milch, 1964); (b) brittle-lacquer methods in studies of a number of skeletal regions (Evans and Goff, 1957); and (c) photostress coating methods in investigations of the dynamics of skull fracture (Gurdjian and Lissner, 1961). I am not aware that Moiré fringe methods have yet been used in biological work, though the method is well known in mechanical strain analysis (Theocaris, 1969) and is being tested in my own laboratory not only for biological strain analysis but also for contouring complex surfaces. A recent technical note by Duncan, Gofton, Sikka, and Talapatra (1970) indicates the apparatus necessary for contouring animal joint surfaces.

Other methods of experimental stress analysis are aimed at obtaining information about mechanical efficiency of form by simulation. These may utilize analogical methods which simulate stress and strain relationships through other physical phenomena. Thus mechanical parameters such as force, displacement, velocity, acceleration, and mass may be modeled in an electrical system by, respectively, current, integral of voltage, voltage, derivative of voltage, and capacitance. Or again, Prandtl's membrane analogy may be used to investigate stresses in a cross-section of irregular shape by investigating the slope of a membrane stretched over a closed supporting boundary with a pressure difference between the two surfaces of the membrane. The slope of the membrane in such a situation is derived mathematically in the same way as the shear stresses at corresponding points in a similarly shaped cross-section of a mechanical shaft under torque.

It is also possible to study stress-strain relationships by utilizing the physical property of photoelasticity. Here the analogy is not with another physical system, but rather is one using a mechanical model; it is this technique that we are currently using. However, it must be emphasized at the onset that this method works in the in vitro situation, through analogy, and involves a series of approximations. One of the chief merits of the method is its visual attractiveness giving as it does an immediate and tangible picture of mechanical efficiency in the structure or detail being investigated (Frocht, 1941; Coker and Filon, 1957).

The technique utilizes the property of reversible birefringence in certain transparent materials when viewed with polarized light under stress. Plastic models of anatomical structures are therefore analyzed. The method involves only approximations, because models can never be replicas of biological objects: thus (a) two-dimensional simplifications are often necessary; (b) only the greatest and most obvious of the external forces acting on an object can be investigated; (c) functional studies have so far been mainly confined to static experiments. These are all serious qualifications, the nature of which should be understood before attempting to make biological deductions from the physical data.

A number of these qualifications can be met by more sophisticated researches. For instance, utilizing frozen stress techniques, it is possible to allow for three-dimensional situations (e.g., Leven, 1955). It is also possible to take into account the anisotropic properties and nonhomogeneous nature of bone by using photoelastic coatings (e.g., Holister, 1961). Dynamic situations can be studied (Flynn, Feder, Gilbert, and Roll, 1962). All these improvements of the method are difficult and time consuming to apply. Nevertheless, such pilot studies as have been carried out so far suggest that the new methods, while they give better answers than simpler methods, do not negate those results; rather, they confirm and extend them. Despite the several sources of doubt and even though the more elaborate techniques may partially eliminate them, the simple photoelastic method remains an excellent tool for the analysis of stress within structures of complex form. The approximations that are required and the errors arising from them are not usually of sufficient magnitude to invalidate general results. The marvelous advantage which the photoelastic method has over theoretical stress determinations is that the former method enables investigation of complicated irregular shapes (as of bone).

Most of the studies that have been carried out to date have related to the demonstration of the mechanical efficiency of biological structure, for instance in the knee and foot (Smith, 1962; Preuschoft, 1970), and a variety of bony structures in both normal and surgically altered cases (Pauwels, 1965, and Kummer, 1959). However, once a reasonable degree of mechanical efficiency is accepted as a big (if not the biggest) element in shaping bony structures, then the technique can be used in a comparative manner in order to evaluate differential biomechanical efficiency.

A second set of comparisons can be introduced as an element of research strategy within the examination of a given shape before its attempted comparison with other biological specimens. That is, for the mechanical arrangement in any given situation, models of increasing complexity (i.e., increasingly more like the true biological situation) can be analyzed in order to provide us with a real feel for the differences produced by successive approximations. Thus the ultimate leap to the true biological situation (which must always be made, but which, nevertheless, is always inferential) is less likely to provide ideas that are grossly misleading.

A Description of Photoelastic Analysis
If a plate of one of the photoelastic plastics is subjected to stress, it is found to behave in the crossed, circular polariscope (figure 72) like a uniaxial crystal, the optic axis of which is parallel to the surface of the plate. As in the case of a natural crystal, the ordinary and the extraordinary light wave produced by the crystal are not transmitted with the same velocities. Thus, when they emerge from the plate, they have a relative path retardation, the result of which means that at some points in the image the rays extinguish one another, giving a series of dark bands, while at other points in the image they summate, giving a series of light bands. These bands are known as isochromatic lines or fringes. The magnitude of the relative

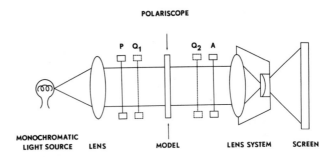

Figure 72. The crossed circular polariscope used for two-dimensional photoelastic stress analysis.

path retardation at any point in the given plate is directly proportional to the differences between the principal stresses and the thickness of the plate. Given that this latter is constant throughout the model, the number of fringes thus reveals the first element in a stress analysis situation—the difference between the principal stresses.

In a field of nonuniform stress the difference between the maximum and minimum principal stresses, and hence the relative path retardation, varies from point to point. Wherever the differences between the maximum and minimum principal stresses have the same value, the retardation is the same; thus on examining a model in monochromatic light we find that dark lines define the loci of points where the retardation is equal to an integral number of wave lengths of the light used. The dark line corresponding to a retardation of one wave length is termed the first-order fringe; the line corresponding to two wave lengths retardation is the second-order fringe, and so on.

Since the maximum shearing stress at any point is equal to half the difference between the maximum and minimum principal stresses, the fringe pattern gives the maximum shearing stress at every point in the model, provided that the stress equivalent of the fringes is known. Also, at a free boundary the principal stresses are respectively parallel and normal to the boundary, and the latter is equal to zero. Therefore the boundary stress at any point on a free edge is equal to the principal stress parallel to the boundary and can be determined directly from the fringe pattern.

It will be seen from the foregoing that the fringe pattern gives an immediate view of the *relative* magnitude of the stresses at every point in the model (figure 73). In many practical cases, either the actual maximum shearing stresses or edge stresses are critical, and the fringe photograph effectively solves the problem.

We have so far assumed in our discussion of stress patterns that the models are examined in circularly polarized light, for instance with a pair of quarter-wave plates interposed between the analyzer and polarizer and appropriately aligned. If the models are examined in plane polarized light (figure 74), and if at any point the direction of one of the principal stresses in the model coincides with the direction of polarization of the incident light, that point will appear dark when viewed through the crossed analyzer. The locus of points at which the principal

Figure 73. The production of fringes during photoelastic analysis. The difference in phase between the ordinary and the extraordinary light rays causes appropriate black and light bands (if monochromatic light is used) or rainbow-colored bands (if white light is used). This feature relates to the relative amounts of stress in the model.

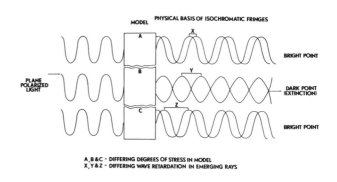

Figure 74. The production of isoclinic bands during photoelastic analysis. In this case light and dark bands are produced by summation and extinction respectively of rays polarized in parallel directions and at right angles. This feature relates to the relative directions of stresses in the model.

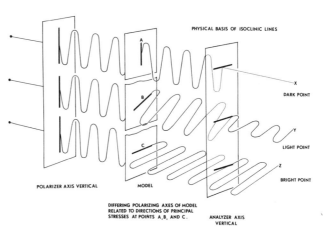

stress direction coincides with the direction of polarization appears as a dark line across the model. Such lines are known as isoclinic lines.

In general the isoclinic lines move across the model as the polarizer and the analyzer (maintaining their crossed positions) are rotated together. In this way, the isoclinics can be differentiated from the isochromatics; these latter remain stationary as the lenses are rotated together, but change as the load on the model is varied, thus producing variations in the values of the maximum and minimum principal stresses. The isoclinic curves are plotted by varying the setting of the polarizer in steps of a few degrees as appropriate, and from these the directions of the principal stresses in every point of the model can be obtained.

It is possible, then, to construct two orthogonal sets of lines (the trajectories) which outline the directions of the two principal stresses at each point in the model. During such a construction, stress trajectories may appear crowded together. Crowding of trajectories gives an impression of regions of high stress values; such an impression, of course, is false; information obtained from isoclinics concerns solely stress directions.

The use of these two properties of the stressed photoelastic plastic allows one to model the gradient of both direction and relative amount of stress within the model.

Functional Significance of Bone Form: Experimental Stress Analysis

Stress Analysis of Bone Shapes

In adapting this photoelastic method to the stress analysis of a bone, drawings of sections of the bone are prepared. Models of these sections are made in photoelastic plastics. It is evident that these models are only approximate reproductions of the original sections. The models are then subjected to external forces which simulate in direction and relative magnitude those acting on the region of the bone included in the original section. Compressive forces are applied directly to regions representing articular surfaces and contacts with the environment. Tensile stresses are applied to extensions of the model representing muscles and ligaments. In both cases the immediate method of application of the load is likely to differ from that of the biological situation. The principle of Saint-Venant suggests, however, that although this has a marked effect upon stress patterns and directions in the locale of loading, it has very little effect on the overall stresses in the main body of the model (Timoshenko, 1955).

For a complete photoelastic analysis we require to know the separate principal stresses and their directions at every point in the model. However, in the present study the forces utilized in the production of the results are the same for each model, and the models are all similarly scaled. As the biological situations which the method imitates are from animals of quite different overall bodily dimensions, where actual forces must differ greatly during life, the real values of the principal stresses are of little interest. What is of interest in this situation is the comparison of stress values from one part of a model to another (a *ratio* of stress values within a single model) and of the subsequent comparison of such ratios from one model to a second. These can be obtained easily by direct comparison of the number and distribution of the isochromatic fringes—thus demonstrating regions of high and low shear stresses for comparison. Because of loading and scaling differences, it is most important not to compare *direct* values from one model to another. Therefore, further studies to obtain absolute values of the principal stresses have not been carried out. However, the materials are calibrated to make sure that the fringe values (amount of stress per fringe) remain the same in each test piece. The strain-optical coefficient is known from the manufacturer's specifications for each plastic.

From the isoclinic lines the pattern of trajectories of the maximum and minimum principal stresses are constructed using a drafting machine. Again, however, the principal use of the data is in pair-wise comparisons of ratios across the shapes of different species. In this way comparisons can be carried out by visual inspection of the resulting trajectorial diagrams.

The Form of Hominoid Hand Bones

A number of studies are now progressing, some of which attempt to test ideas that have resulted from prior knowledge of function and anatomy, others of which are exploratory and suggest areas where more classical studies might proceed. One anatomical region currently under investigation is the digital ray of the great apes and man. Recent investigations (Tuttle, 1969) have outlined a mechanism and some associated anatomical features of knuckle-walking in African apes. The current study attempts to test and amplify these anatomical inferences utilizing the technique of photoelastic analysis. The relationship between a knuckle-walking simulation and various architectural features of digital rays has been examined, including such subjects as the relative lengths and widths of metacarpals and phalanges, the nature of their longitudinal curvatures, the disposition and shape of their bony ridges, and the orientations of their joint surfaces (figure 75).

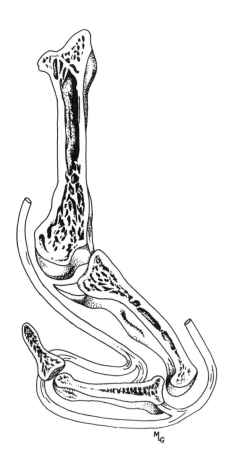

Figure 75. A diagrammatic longitudinal section through the principal digital ray of the chimpanzee when placed in a typical knuckle-walking posture. Anatomical features are simplified and slightly exploded.

The studies have been carried out as a series of paired comparisons between individual features of chimpanzees and orangutans so that deductions will be comparable for both forms. In this way, though the mechanical efficiency of neither form can be determined absolutely, relative differences in mechanical efficiency between them can be defined (Oxnard, 1969a; 1972a).

The particular specimens chosen are adult, wild shot, nonarthritic individuals. From these are prepared a series of diagrams of the individual bony elements. The diagrams are then used to obtain a graded series of models (a) utilizing only

Functional Significance of Bone Form: Experimental Stress Analysis

information about the lengths and widths of the elements, (b) incorporating in addition the longitudinal curvatures of the elements, and (c) adding the details of the bony buttresses evident in the sections. Further amplifications of the models are made so as to introduce information (d) about joints between elements and (e) about major tendons attached to elements.

The results of this part of the study (figures 76 and 77) show the distributions of the shear stresses in the models of the chimpanzee and orangutan during a simulated knuckle-walking posture.[1] First we may study the analysis for the chimpanzee alone (figure 76). Although a series of metrical comparisons can be made on data, the results are so obvious that simple inspection of the pictures provides the major conclusions. As one would expect, the most simple analogy of all suggests that those elements (length and width alone) of information about the chimpanzee digital ray provided by the model are of very poor mechanical efficiency (top left, figure 76). But the progressive series of pictures (top row, figure 76) shows clearly how successive refinements of the mechanical situation provide more and more mechanically efficient shapes within the knuckle-walking posture. By far the best refinements are provided, of course, when joints and tendons are simulated (bottom row, figure 76).

An examination of the series of pictures for the orangutan (figure 77) confirms that the same is true in this case. But the real test for mechanical efficiency in knuckle-walking is the cross comparison between the two sets of pictures. This makes it clear that the degree of efficiency of the chimpanzee is considerably greater than is that of the orangutan. And the same conclusion is obtained whether one examines a very simple analogy (top row, figure 77) or the better analogies (bottom row, figure 77). We may assume that we are therefore correct in making the suggestion that if we could model the true anatomy, the chimpanzee would still be more efficient within this context.

This result is hardly surprising, for we know that the chimpanzee is capable of knuckle-walking and the orangutan is not. But would the result have been of interest if the orangutan shapes had actually been those of some unknown fossil form? Would the data have suggested inability in knuckle-walking in such a case?

A similar series of comparisons has been carried out utilizing again the orangutan and the chimpanzee, but in this case the study has examined the digital ray in a posture that corresponds more to arm-hanging-climbing contrasts (figure 78). Rather than describing again the full series of models (straight, curved, irregular, one joint and one tendon, a different joint and two tendons), let us simply look at single pictures from the middle of the two series compared directly. Figure 79 shows the result for the chimpanzee and the orangutan when the shapes that are compared contain information about dimensions, curvatures, and irregularities only. And as in the previous case, the result is equally obvious: here the orangutan is considerably more efficient than is the chimpanzee. But, based upon the

---

1. This analysis was carried out for the orangutan notwithstanding the fact that the animal is not adept at knuckle-walking. For the technique to be viable it must be applicable to the unknown (fossils) for which we may know little or nothing of true habits.

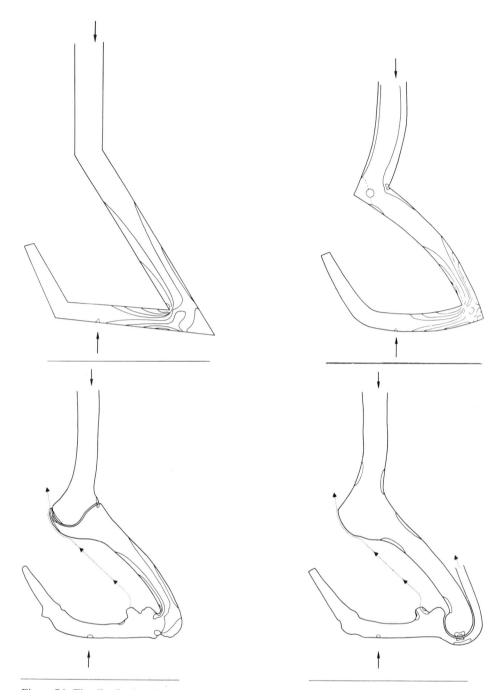

Figure 76. The distribution of isochromatic fringes in the elements of the finger in the chimpanzee in a knuckle-walking simulation. The first picture includes information about lengths and widths alone; the second adds information about curvatures; the third, about bony ridges and stops. In the bottom two pictures the simulation has been much improved by including tendons and joints.

The greatest efficiency is seen in the last picture, where the stresses are nowhere greater than a single fringe: i.e., the distribution of the stresses is almost equal throughout the model. This is somewhat less the case in the next to last model, in which the proximal phalanx has three fringes as compared with other parts of that model. Both of the last two models are much more efficient than the upper three. Efficiency has been increased throughout the series by the anatomical improvements.

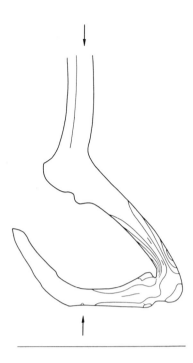

quantitative aspects of the results, the degree of inefficiency of the chimpanzee for the hanging simulation is not nearly as great as is the degree of inefficiency of the orangutan for knuckle-walking. And this accords well with the known behaviors of the two: chimpanzees do hang, although not so well as orangutans; orangutans scarcely knuckle-walk at all. Again this might seem a very obvious study. But I think we can see the value of the technique if we assume that either of these animals had been an unknown fossil form.

Similar studies have been carried out for man and the gorilla. The findings for the gorilla more or less replicate those for the chimpanzee; both animals are highly efficient within the knuckle-walking context, and mildly efficient within the arm-hanging-climbing context. Man is grossly inefficient in both functional roles (figure 80).

The more detailed results of the above studies supply considerable information as to which aspects of shape contribute to the mechanical differences between the different forms. They give clear indication of the facets of shape that might be defined in metrical studies of the sort previously discussed; such metrical studies allow a more sophisticated examination of the variation within groups that is not possible utilizing the photoelastic method. However, photoelastic studies are especially helpful in forming a basis for assessing the functional meanings of particular architectural features not only in extant forms but also in fossils.

The technique can thus be seen to be of value in functional morphological studies at several different stages. In the case of studies on the shoulder girdle, the

132 Form and Pattern in Human Evolution

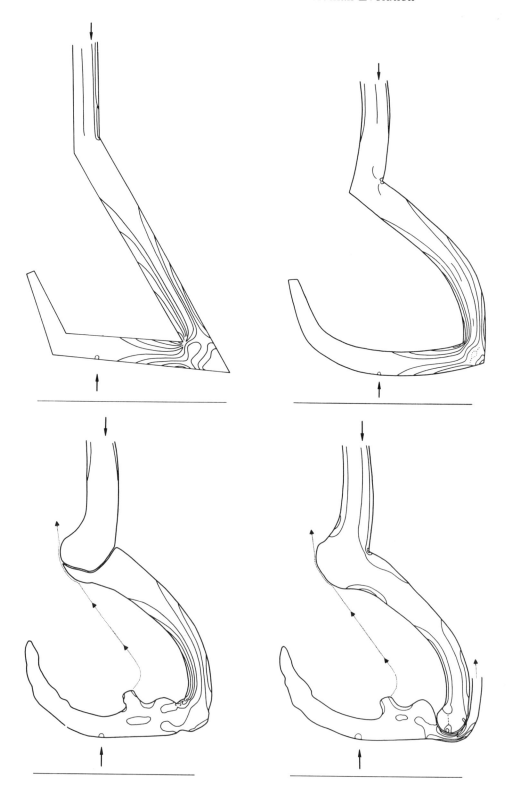

Functional Significance of Bone Form: Experimental Stress Analysis

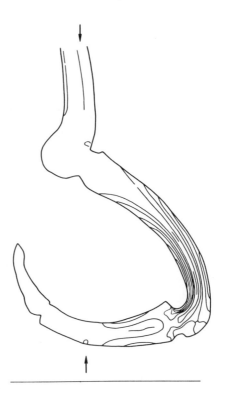

Figure 77. A comparison of the isochromatics in simulated knuckle-walking in the orangutan. The same series of anatomical changes have been made in the respective models as in figure 76.

As in the case of the chimpanzee simulation, the same series of improvements in the models has produced a series of improvements in the distribution of the isochromatic fringes in each model; the fourth and especially the fifth models are more efficient than are the others. But the cross-comparison of the extent of the relative efficiencies of the different models between figures 76 and 77 indicates clearly the much greater efficiency of the models in figure 76 (chimpanzee) in the knuckle-walking simulation than in this figure (orangutan).

method allows an independent means of testing the validity of functional inferences previously made on the basis of the mathematical treatment of extensive data. Thus the various studies of chapter 3 have suggested that one set of differences that can be observed between different primate scapulae associates a "craniolateral twist" of the bone with "increasing capability for scapular rotation, as in arm-raising." Accordingly, a series of photoelastic models of the scapula are examined in order to test whether, truly, a change in shape of this type could be related to mechanical efficiency in such an anatomical situation. The results are summarized in figure 81. Here a scapular shape equivalent to that of a macaque is clearly more efficient for bearing compressive forces than is that of a gibbon. This relates well to the notion that the scapula of a macaque bears such stresses during its normal locomotion and that of a gibbon does not. When, however, there is added to the compressive forces acting upon the model of the gibbon scapula a set of rotatory forces such as would obtain from the actions of the upper part of trapezius muscle and the lower digitations of serratus magnus muscle during arm-raising, the result is a more efficient mechanical system for the gibbon. This provides independent confirmation of the above ideas.

In the case of the studies on the digital ray just described (a situation in many ways biomechanically more complex than the shoulder), the technique is being used at an early stage in the studies in order to help define those features of the bones which are likely to be relevant to function, and which may then be defined and measured for examination by multivariate statistical and other methods. At

Figure 78. A diagrammatic longitudinal section through the principal digital ray of the orangutan when placed in a typical hanging posture. Anatomical features are simplified and slightly exploded.

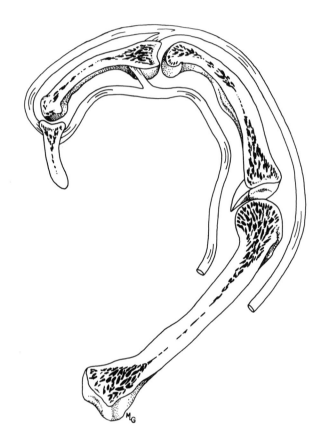

the same time, however, it is clear that this particular study also allows the possibility of directly examining information from fossil remnants. For instance, Napier (1962) shows a photograph of a drawing of a phalanx from the Olduvai hand; this is a well-curved element; it is possible to study the efficiency of this bone in each of the mechanical situations known to exist in modern species. Of course, such a study cannot define what that shape was capable of doing in a biomechanical sense; but it can provide evidence about those mechanical functions for which it is clearly inefficiently designed. This may give useful information in speculations about possible functions for that particular fossil.

It may even be possible, on occasion, to go further. Supposing a fossil creature were, in fact, carrying out some function that is unique, this may be revealed by the technique. For just as engineers can change a shape to fit a particular set of forces (the essentials of designing by photoelasticity), so it is also possible for us to change a set of forces to fit a particular shape. Such definitive studies have not yet been carried out on the Olduvai hand bones primarily because access to the original materials and good casts is necessary. However, in lieu, and mainly to demonstrate the nature of the scientific strategy, preliminary investigations have

Functional Significance of Bone Form: Experimental Stress Analysis

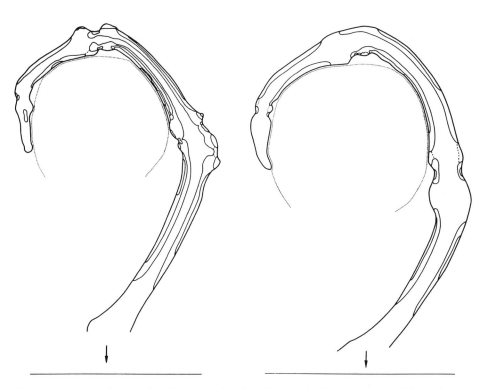

Figure 79. The distribution of the isochromatic fringes in models of the orangutan (right) and chimpanzee (left) when placed in a hanging posture. In this case only the middle model of the series is shown (displaying information about dimensions, curvatures, and bony buttresses). Both models are efficient, though that for the orangutan is considerably more so. The same information is also given by the more complicated analyses (not shown), in which joints and tendons are included, and in which the loading more nearly approximates to that obtaining during life.

been attempted utilizing photographs of the proximal phalanx. These are reported in chapter 7.

Finally, a third way in which this technique may be applied is evident from a series of studies that is currently being carried out upon the mechanical efficiency of the pelvis. In this case the studies are proceeding in step with the osteometric investigations described above; to date, work has scarcely progressed far enough to supply even tentative data. We can see, however, that suggestions derived from the canonical analysis of the pelvic dimensions (for instance that the various osteological features within the pelvis appear to congregate in subsets that seem to have functional meaning within locomotion; see chapters 3 and 7) are exactly capable of being tested by the photoelastic analogy.

Theoretical Applications: Adaptation of Bone to Stress

Photoelastic analysis can be utilized within morphology for a further set of problems: those relating to theoretical aspects of the shape of skeletal elements.

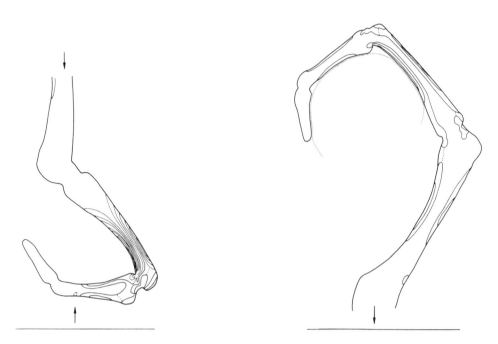

Figure 80. The distribution of isochromatic fringes in single knuckle-walking and hanging simulations for the elements of the human digital ray. The inefficiency of these shapes, especially in the case of the proximal phalanx, is most marked. The same picture obtains for more complex models.

Thus recent studies of the mechanism of adaptation of bone to stress have suggested either that the role of tension in bone is of little importance to bone adaptation as compared with compression (Frost, 1964), or, if the role of tension loading is extensive, that current hypotheses as to the mechanism of bone adaptation need improvement (Currey, 1968). Clearly many parts of bones must be under considerable tensile stresses at different times during function. Any element that undergoes bending, for instance, will experience tension down one side as severe as the compression down the other. However, if such an element is bent first one way and then the other, it is likely that over any given time period the net effect will not be any large degree of tension. If at the same time other compressive forces are acting (e.g., due to weight-bearing), then such an element is liable to be under an algebraic, vectorial and temporal totality of compression.

But there are a number of regions of the bony skeleton where it might be thought, intuitively, that tension over a period of time might indeed be the more powerful and frequent state. For instance, such regions as the sites of attachments of muscles must surely bear tensile forces for considerable periods. At the attachments of extremely powerful muscular elements one can usually find anatomical structures (such as intraosseous collagenous fibers at the tibial tubercle, the site of attachment of patellar tendon) that are clear adaptations to the tension which

Functional Significance of Bone Form: Experimental Stress Analysis

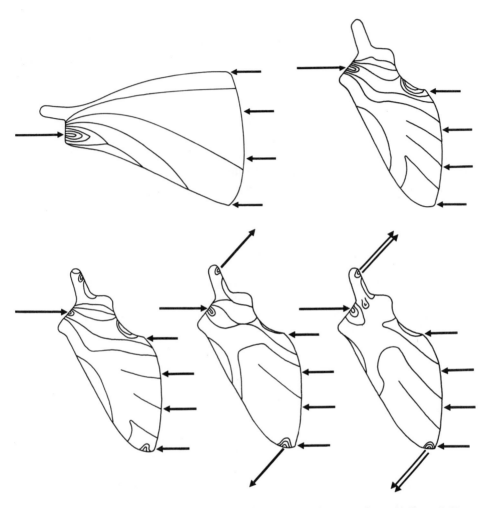

Figure 81. Isochromatics in the scapula. The top picture shows the comparison of baboon (left) and gibbon (right) scapular shapes when loaded compressively as though both animals are quadrupeds. The gibbon appears relatively less efficient in this mode than the baboon. The lower picture shows comparisons of three gibbon scapular shapes when additional rotatory forces are added to the same compressive loads. In the left figure the rotatory stress is zero; in the middle figure it is fairly big; in the right figure it is twice as big again. The larger the rotatory force, the more efficient the gibbon scapular shape.

must exist. However, in all probability the presence of these structures results in the tension's being borne by the collagenous bundles rather than by the bone itself. But there are other cases where the local structure does not appear to be so modified, yet where the anatomical arrangement is such that we may be justified in suspecting that tensile forces are present.

One example is the bone within a crest which serves as the site of attachment of muscular elements pulling in opposite directions across the crest. A second is that within a thin bony plate where muscles attach on each side of the plate and surely produce tensile forces within the plate. A third is within a sesamoid bone where the tendon attached to each end may be expected to produce an overall state of tension within the bone.

The photoelastic method is able to model these three different situations in order to determine if indeed tension or compression exists as the principal stress feature in these structures (Oxnard, 1971). The first example is that modeled by the crest of the spine of the scapula (figure 82), to which are attached deltoideus muscle on the one side and trapezius muscle on the other. In most animals these muscles act alternately because the former is a retractor of the quadrupedal limb (power stroke), while the latter participates in protraction (return stroke). The net result of such a situation is alternate bending, which does not give rise to predominance of tension in the scapular spine (figure 83). Only if the two muscles frequently contract together will the crest of the spine be under tension for considerable periods of time. It so happens that the anatomical situation in some bats is such that (pending proof from electromyographical recordings) these muscles probably do cocontract during flight. Certainly both are situated so as to perform the same action (dorsal lifting of the wing) rather than opposite actions as in most quadrupedal animals. Photoelastic analysis suggests that in this case the crest of the spine is genuinely under tension (figure 84).

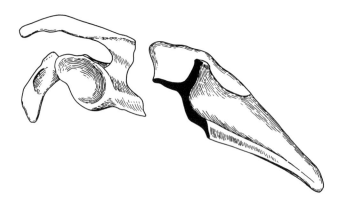

Figure 82. A cross section of the blade and spine of the scapula to show the shape of the crest of the spine in most mammals.

It is of interest, therefore, that in these bats there is no bony crest to the scapular spine: this region consists of a collagenous sheet, bounded by a fairly powerful ligament, from which arise the two cocontracting muscles (figure 85).

The second anatomical situation—that of muscles arising from either side of a bony plate—has also been modeled with the technique (figures 86 and 87). The blade of the scapula is a sheet of bone with fairly powerful muscles, infraspinatus and subscapularis, attaching on each side. When these are the only muscles acting, that region of the bone will be under tension during their cocontraction (in this example it is clear that they do cocontract most of the time, because, though

Functional Significance of Bone Form: Experimental Stress Analysis

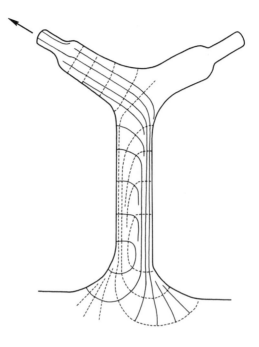

Figure 83. The stress trajectories in a model of the crest of the scapular spine when subjected to bending as by contraction of the muscle arising from one edge. The model is subject to compression on one side and tension on the other.

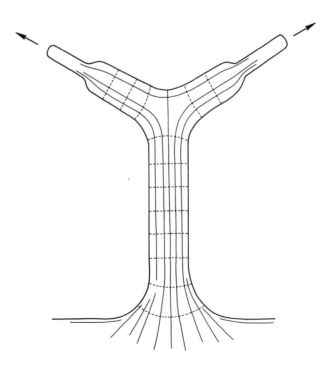

Figure 84. The stress trajectories in a model of the crest of the scapular spine when both cristal muscles contract together. Consideration of edge stresses show that tension is present.

Figure 85. A section of the scapula to show the spine and crest in a bat, an animal where both cristal muscles co-contract during flying.

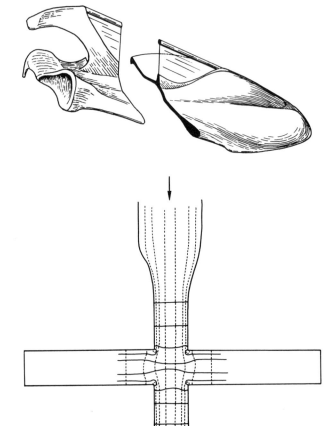

Figure 86. The stress trajectories in the scapular blade during compression along the length of the blade. The vertical limb represents a cross section of the scapular blade, thick peripherally, thin centrally; the horizontal arms allow for the addition of tensile forces at right angles to the blade as during contraction of the muscles attached to the scapular fossae. In this simulation, these muscles are not active; consideration of edge stresses shows that compression is present in the scapular blade.

they have different individual actions on the shoulder joint, they are both stabilizers of that joint and this is their most important action). But in the majority of quadrupedal animals, the forces due to these two muscles are not the only ones acting on the scapular blade; it must also take up the extensive compressive forces due to the quadrupedal stance. Those forms which are most likely to have tension existing in the blade of the scapula are the ones in which the two muscles are absolutely most powerful (i.e., biggest), and those in which the scapular blade is dorsal in position so as not to take heavy compressive stresses in weight-bearing (rather than lateral, as is the case in quadrupedal animals). Such forms comprise

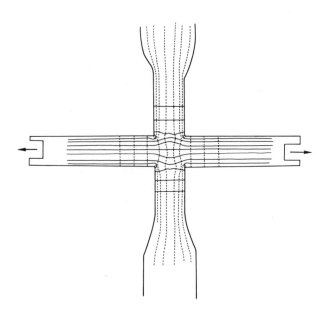

Figure 87. The stress trajectories in the scapular blade modeled as in figure 86 but with tensile stresses added to simulate the muscles of the scapular fossae. Consideration of the edge stresses in this case shows that net tension could exist in the center of the scapular blade if the compressive forces are reduced enough.

the great apes and man. It is of considerable interest, therefore, that, of all primates, these are the ones with the greatest relative thinning (and almost the greatest absolute thinning) of the scapular blade. And it is among these species that a foramen bridged by a collagenous sheet (from which arise the two muscles) may be found in occasional specimens (figure 88). A likely hypothesis must be that the anatomical situation is such that the scapular blade is here under small compressive forces only, and that occasionally they may be reduced to the point where the net result may be tension.

Both of these examples (the scapular spine and scapular blade) are where bone is normally present and where the photoelastic technique suggests that forces are compressive. In both cases, however, the analysis suggests that sometimes tension may predominate, and this coincides with the finding that bone may not be present; equivalent collagenous structures may occur.

The third example, a sesamoid bone within a tendon (figure 89), is a case where intuitively we would guess that there should exist considerable tension. And in the simple case in which the only forces that act are those due to tendinous attachments, the technique shows easily that tension would indeed be found (figure 90). But when compressive stresses due to the articulation of the patella with the femur (in this example the knee was examined) are included, the rather surprising finding is that the great bulk of the sesamoid is under compression (figure 91). Only the thinnest rim of material on the outside is under tension; this material is usually composed of coarse-bundled bone within which are extensive collagen fibers that are related to the attached tendons. The whole external surface is also usually reinforced by a tension-resisting, collagen-strengthened membrane, again related to the two tendons. Clearly, similar areas of overall compression will exist in any

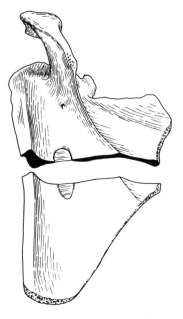

Figure 88. A cross section of a scapula of a gorilla in which the central bone is absent. Present is a collagenous sheet formed from the periosteal membranes of the two sides of the scapula. The muscles of the scapular fossae originate in part from these membranes (from a photograph by T. Tonkinson).

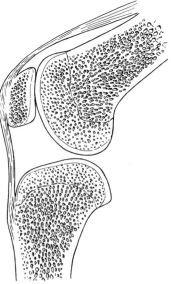

Figure 89. A diagrammatic longitudinal section of the human knee joint in partial flexion.

tendon winding around a skeletal element; but compression will only be a predominating factor in any given single region of the tendon in cases where the length of contact is equal to or greater than the amount of movement of the tendon. These are the cases in which sesamoids are found. In others (e.g., the curvature of the obturator internus tendon around the ischium) compression does not predominate in any one region, and bone is absent.

Functional Significance of Bone Form: Experimental Stress Analysis

Figure 90. The theoretical but incorrect pattern of stress trajectories in a sesamoid bone under tension alone. Consideration of edge stresses shows that the entire sesamoid is in tension.

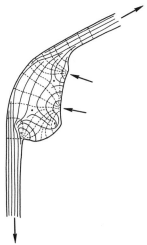

Figure 91. The pattern of stress trajectories in a sesamoid under a more natural load. Consideration of edge stresses suggests that the major portion of the bone is under compression. The only part under tension is a thin rim along the outer surface.

These ideas are the result of pilot studies. Researches are presently in progress to determine to what extent the results can be generalized. For instance, although matters seem fairly clear in the above macroscopic examples, this seems less obviously the case when we examine microscopic instances, such as the insertion into a tubercle or pit of a strong tendon, the attachment of a tooth within alveolar bone, or the intratendinous bone found within bird limbs. These microscopic situations are more complicated, although even here initial studies suggest that bone elements still bear a totality of compressive forces, with collagen fibers the resultant tensile ones.

Yet the three cases examined to date all point to the same conclusion: that predominantly compressive forces are necessary for bone to remain. It tends to confirm Frost's (1964) opinion that the role of tension-loading in bone may have been exaggerated in biomechanical thinking. The concept is also supported by the recent work of Justus and Luft (1970), who have shown increased solubility in individual hydroxyapatite crystals under tension. This is a conclusion clearly of some importance to those interested in the investigation of the form of bones in extant primates and in the application of the results to the shape of fossils, both of which studies comprise a necessary background towards understanding the evolution of the group.

7 Extrapolation to "Unknown" Data

Introduction

Nearly all of the techniques described in previous chapters can be used in one way or another for the examination of information from unknown or fossil organisms. When data from a fossil primate are utilized, the resultant speculation may have greater or lesser likelihood of being correct according to the extent to which it relates to data from extant forms. No further test of such speculation is available unless new data are utilized (new data from the same specimens, similar data on new specimens, or both). Though it is not possible to make direct tests without such new data, it is possible to make tests in kind using extant forms and pretending that they are unknown, that they have missing data, that they have been reconstructed, and so forth. Such tests may be useful to indicate the general nature of pitfalls.

A Test with a "Living" Unknown

A test of this type has now been done for the seventeen-dimensional data on the shoulder girdle utilizing *Daubentonia* as a test piece.[1] Two analyses are available: one with *Daubentonia* excluded from the main study but entered later using the loading factors initially obtained, and a second with *Daubentonia* included in the main study so that it makes its own contributions to the derivation of the loading factors. The first of these techniques is what would normally be done if *Daubentonia* were a fossil; the second method is that applied considering the genus as extant. Interrelationships of groups are investigated at the superfamily level; this is done simply to keep the number of groups small for the purposes of the test.

When *Daubentonia* is entered indirectly after the main analysis, the genus falls among the other superfamilies lying at its own locus in the same part of the canonical space (figure 92). To those who know anatomy of the shoulder of *Daubentonia*, this is a curious result, because though *Daubentonia* is currently recognized as a primate (i.e., in a taxonomic sense it belongs with the other primates), the shape of its scapula and clavicle differ greatly from those of any other member of this order. The scapula, for instance, is markedly twisted in a plane at right angles to the general plane of the scapular fossae. This difference from other primates is so great as to suggest that *Daubentonia* is uniquely different from them in this respect, a conclusion opposite to that given by the indirect canonical analysis (though it is true that no direct measure of this anatomical feature is among the seventeen dimensions analyzed).

When, however, *Daubentonia* is incorporated directly into the analysis from the beginning, so that the variation and covariation of its own dimensions can make their contributions to the separations, then the results are as seen in figure 93. *Daubentonia* lies clearly outside the canonical space occupied by the other primates; this result is in accord with what we know about the form of the scapula

---

1. Ashton, Healy, Oxnard, and Spence (1965) did calculate the position of *Daubentonia* in the earlier study of nine shoulder dimensions but without taking close notice of possible indications that the analysis might be incorrect.

Figure 92. The canonical separation of the superfamilies of the primates in the examination of the seventeen-dimensional shoulder data. The bivariate plot of canonical axes 1 and 2 is shown. The marker represents two standard deviation units. The position of *Daubentonia*, entered indirectly, is shown. It falls close to the other forms, especially to the other Prosimii.

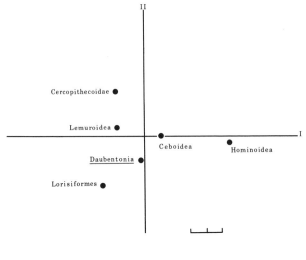

Figure 93. The difference between this and the preceding figure is that *Daubentonia* has been entered into the analysis directly so that it may make its contributions to the derivation of the loading factors. The diagram shows that *Daubentonia* is not close to the other superfamilies: the information supplied by the previous picture is spurious.

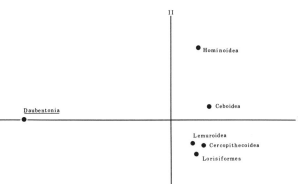

in this animal, even though again no direct measure of the main difference — scapular twist — is included in the analysis. (The functional associations of these features of the shoulder girdle of *Daubentonia* are still wanting; their definition rests upon the determination of some facts about the use of the shoulder in this rare and curious animal). The result therefore shows that it may be a doubtful procedure to investigate an unknown form by merely spotting it indirectly into a previous canonical analysis. This is, in fact, what has been attempted by a number of workers using multivariate techniques to examine fossil data.

An important question now arises: Given the recognition of the misleading results, can there be discerned in an indirect study any information to foretell its misleading nature? Curiously enough, the hindsight of this investigation suggests that there can. Careful examination of all the canonical variates may provide the needed evidence. For instance, in the indirect study the placement of *Daubentonia* in the higher canonical axes (which are rarely examined closely once it is determined that they contain little information about the separation of the groups) shows that the new form lies at the edge of the distribution of genera along several higher axes. Clearly, although there must always be genera lying at the edges of such separations, we do not expect the same genus to occupy such a position in a number of axes. For if that were the case, then the information relating to that genus would, in a direct analysis, have been brought forward into earlier variates; this is demanded by the nature of canonical analysis, where the maximum separations must occur in the earliest axes. However, for a genus that is entered indirectly afterwards, this anomalous result might well occur. Such a finding therefore suggests that the position for *Daubentonia* in the indirect analysis is indeed aberrant. The direct analysis confirms this and gives further information as to the large degree of the aberration.

A second way of obtaining this result is to utilize the information hidden in the generalized distances (in the full seventeen-canonical space) by performing a cluster analysis on those values. This technique clearly recognizes when the result obtained from the examination of earlier canonical axes differs from the overall generalized distances. The demonstration of this is supplied by examining the differences between the minimum spanning tree as derived from the canonical two-space and from the full canonical seventeen-space (generalized distances). This can be achieved by plotting the two minimum spanning trees on the same diagram (figures 94 and 95). Thus figure 94 demonstrates that the information given by the canonical two-space in the direct analysis is qualitatively similar to that obtained from the full generalized distances. In contrast, figure 95 shows that in the indirect analysis this is true for the main superfamilies but false for

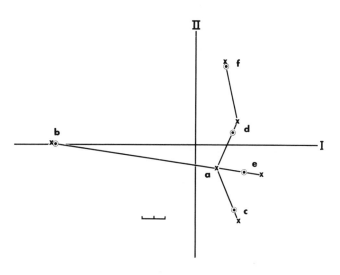

Figure 94. Minimum spanning tree superimposed upon canonical analysis of figure 93. In this figure the dots represent the positions of the primate superfamilies in the bivariate plot of canonical axes 1 and 2; the crosses represent additional distance due to the generalized distances in the canonical seventeen-space; b = *Daubentonia*; a and c–f = primate superfamilies. The groups are joined by the minimum spanning tree. The diagram shows that the information within the canonical two-space is virtually identical to that in the canonical seventeen-space: in particular, it confirms the aberrant position of *Daubentonia*.

148   Form and Pattern in Human Evolution

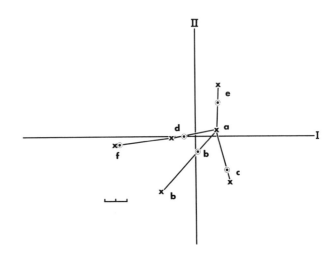

Figure 95. Minimum spanning tree superimposed upon the canonical analysis of figure 92. The symbols are as in figure 94. In this case, however, *Daubentonia* has been entered indirectly. The diagram shows that the information within the canonical two-space is virtually identical to that within the canonical seventeen-space for the superfamilies of the primates (the dot for each superfamily is not too far distant from each respective cross). But for the genus *Daubentonia* (b), the position in the canonical two-space is widely different from that in the seventeen-space. This shows that its position in the canonical two-space is spurious.

the interpolated genus *Daubentonia*. Figure 95 thus supplies the information that suggests that the indirect analysis is incorrect.

It must be emphasized that in both of these illustrations the biological situation has been deliberately simplified by grouping at the superfamily level in order to make the point. Full examination of the entire data set at the generic level satisfies that similar conclusions apply to the canonical analysis (figures 96 and 97); in other words, the pooling of the groups of genera has not produced a spurious result. Furthermore, the cluster analytic result, when based upon the positions of genera rather than of superfamilies, also demonstrates that, in truth, *Daubentonia* is a markedly outlying genus (see chapter 5, figure 66).

These considerations permitted investigation (Oxnard, 1968) of locomotor adaptations in the shoulder girdles of a number of mammals, and formed the basis of rejection of nonarboreal forms from interpretation. Of course, in that study these questions are so obvious (the difference between arboreal and nonarboreal mammals is great) as to require no detailed examination.

A further point then becomes appropriate in studies of this type: Given that *Daubentonia* really does lie outside that part of the canonical space inhabited by the other shapes, does the direct inclusion of the new genus affect the interrelationships among the other forms themselves? Again careful examination of the canonical axes shows that it does not. Although the first canonical axis separates *Daubentonia* from all else in the direct analysis, the later axes separate only the remaining primates among themselves. The effectiveness by which this is done is such that the later canonical variates for each genus differ by less than a second decimal place from those obtained from the study when *Daubentonia* is excluded. This can be seen from the next two figures (98 and 99); axes one, two, and three

Figure 96. Bivariate plot of canonical axes 1 and 2 for the indirect superfamily study when the individual genera are entered. The large symbol in each case represents the position for the superfamily, and the smaller symbols are for the individual genera. The diagram shows that the superfamilies represent reasonable pooled groups of genera.

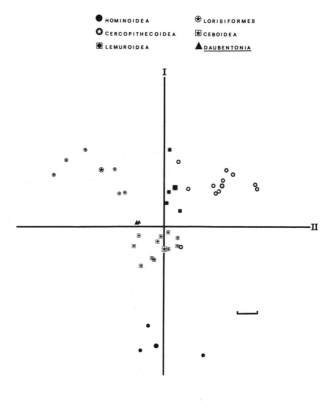

Figure 97. Shows that the result for figure 96 is also true for the direct study. Symbols as in figure 96.

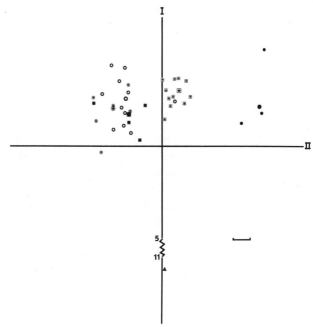

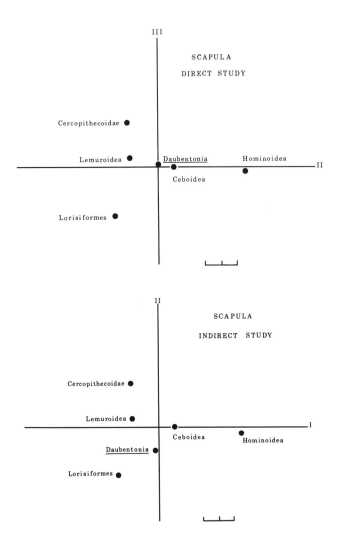

Figure 98. Comparison of direct and indirect scapular studies. Except for the position of *Daubentonia*, the structure of the groups is similar for the direct and indirect studies in the bivariate plots of canonical axes 2 and 3 (direct) and 1 and 2 (indirect).

of the original indirect study are seen to be virtually identical to axes two, three, and four of the direct study (in which axis one separates only *Daubentonia*).

The significance of this feature may be of considerable importance. It suggests that the inclusion of aberrant data may not necessarily affect the relationships of the main forms in the analysis. In other words, the information in the early canonical axes for the main bulk of primate forms is robust and appears even when aberrant material is included in the study. In like manner one would hope that the inclusion of fossil data would show whether or not the fossil was part of the general data space that was being examined. The robusticity of the information contained within these earlier canonical variates has also been well demonstrated by the fact that it appears even when we examine different subsets of the data

Figure 99. Further comparison of direct and indirect scapular studies. Except for the position of *Daubentonia*, the structure of the groups is similar for the direct and indirect studies in the bivariate plots of canonical axes 2 and 4 (direct) and 1 and 3 (indirect).

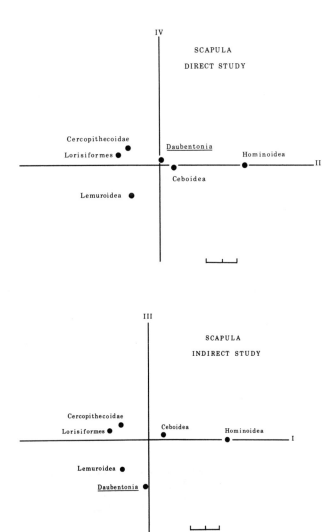

(cf. the results of the locomotor and residual studies with the locomotor plus residual analysis, chapter 3).

Some Fossil Unknowns: the Sterkfontein Innominate Bone

Having performed a test of this nature, it is then possible to look at an example where a genuine fossil (the innominate bone from Sterkfontein—albeit somewhat reconstructed) is superimposed upon extensive data from extant forms. These data are the nine dimensions of the primate pelvis examined individually and reported in a preliminary manner by Oxnard (1966) and Zuckerman, Ashton, Oxnard, and Spence (1967). A preliminary description of the canonical analysis is provided by Zuckerman (1970). The full canonical analysis will be published shortly. In this case, canonical analysis reveals the unique separation of man from all nonhuman primates—a conclusion readily apparent to the eye and already provided by univariate analysis (chapter 2). At the same time, however, considerable differences are found among the various nonhuman primates by multivariate analysis, and these may be readily related to functional differences (chapter 3). The interpolation of the Sterkfontein fossil suggests that it lies approximately halfway between man and the great apes. Like man, it is uniquely separate from nonhuman primates, yet it is far from being identical to man.

Investigations are currently in progress to determine precisely how the fossil differs from man on the one hand and the great apes on the other. Already it is clear that part of the distinction seems to reside in differences in the overall size of the specimens—allometry of sorts seems to be involved.

But it also seems that subelements of the innominate bone can be readily defined. For instance, in a subsidiary canonical analysis of all those dimensions which can be most closely related to muscle function, the fossil is *scarcely distinguishable from the great apes*. In a canonical analysis of those parameters which seem most to be measures related to the positions of joint surfaces, the fossil *resembles man*.

Does this suggest that, in the evolution of upright posture, the actual positions of the joints in the pelvis in relation to the center of gravity change first? The animal is presumably adapting to the upright posture. At this stage muscles still resemble the pattern of apelike forebears. Adaptation to bipedal locomotion superimposed on upright posture appears to be not yet fully developed. We may ask, "what is the nature of this partially developed bipedality?" However, the full details of this study (involving collaborative work between Zuckerman, Ashton, Flinn, Oxnard, and Spence) are yet to be worked out.

The investigations of the Sterkfontein innominate that have previously been carried out (e.g., Le Gros Clark, 1964) have aimed at characterizing the manlike nature of the bone. This is revealed most clearly in the almost entirely human shape of the iliac blade; this feature has been greatly emphasized in portraying the different bones by orienting them in the plane of the ilium. This is demonstrated in plate 3, in which the Wenner-Gren casts of the human, chimpanzee, Pygmy, and australopithecine innominates are viewed in this way. In contrast, though the features of the fossil in which it resembles some apes and monkeys have been fairly well documented in the literature, much less attempt has been made to suggest what might be their possible biomechanical significance in relation to the evolution of bipedalism and the upright posture. This second group of features is well seen in plate 4, in which the same four pelves are shown, but in which case all are oriented as seen in the plane of the ischium. Both of these orientations are arbitrary. Somewhat more realistic might be views of the bone in

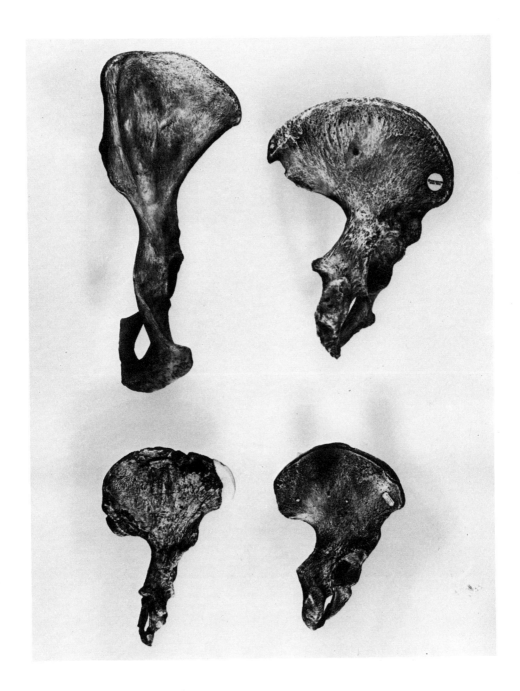

Plate 3. The Wenner-Gren Foundation innominate casts of the chimpanzee (upper left), man (upper right), Sterkfontein fossil (lower left), and pygmy (lower right), as seen when the bones are oriented in the plane of the iliac blade.

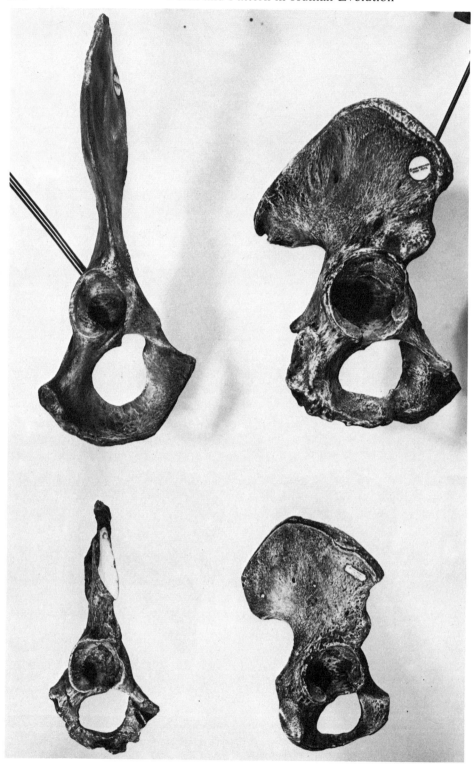

Plate 4. The same casts as in plate 3, oriented in the plane of the ischio-pubic rami.

a direct lateral orientation and a direct anterior aspect. These latter suggest shortcomings in assessments based upon specific ad hoc orientations. It seems clear that canonical analysis has succeeded in objectively pinpointing some of these discrepancies.[2]

Fossil Fragments of Shoulder Girdle

In another investigation, data from fossil remnants became available after completion of work on extant forms (Oxnard, 1968b,c). These data, from the Sterkfontein scapula and the Olduvai clavicle, are extremely restricted and somewhat dubious in nature. Nevertheless, it seems reasonable to attempt interpolation into the extensive shoulder study already performed. Because the data are incomplete, the extent of the interpolation is less than for the innominate bone. Accordingly a series of rather more speculative procedures is attempted (Oxnard, 1969d).

The original data from extant forms allows the examination of the position, in the canonical space, of man in relation to those of all other primates. This procedure suggests that the nearest morphological analogues are the various arboreal forms. Thus the morphological distances between man and, in turn, different *arboreal* forms (saki and uakari monkeys, spectacled langurs and snubnosed langurs, spider monkeys and orangutans) are much shorter than those between man and various *terrestrial* species (patas monkeys and baboons, hanuman langurs, gorillas and chimpanzees) (figure 100). These morphological distances are not to be confused with concepts of genetic or phylogenetic distance. In other words, it may be *morphologically* easier to derive a shoulder like that of man from the shoulder of some arboreal creature (not from any of the above—indeed, if molecular data mean anything, from a progenitor genetically related to the African great apes, but at a time before these latter had become morphologically adapted to terrestrial locomotion) rather than from the shoulder of a terrestrial form.[3]

With this information as background, the interpolation of the fragmentary data from the Sterkfontein scapula (referred to *Australopithecus*) and from the Olduvai clavicle (originally referred to *"Homo habilis"*) may then be carried out. These two fragments may, at this level of investigation, be reasonably lumped into a

2. Canonical analysis still suffers from the defect that it is necessary to define points for measurement in order to use this method. The definition of these points may well be a source of legitimate argument among morphologists. Studies outlined in chapter 8 suggest some of the ways in which a bone may be examined without the subjectivity of defining points. Such preliminary studies are utilizing the technique of the medial axis transformation (see chapters 1 and 8). The method is capable of adapting information from both sources, that is: of defining a shape without utilizing reference points and then of including information about reference points if desired, within the analysis.
3. This seems to be supported by the findings of Lewis (1971) in a study of the hominoid wrist. He concludes, "In spite of the lack of precise indications as to the time of origin of the hominid line, there now seem to be adequate morphological grounds for a return to the view that man and the African great apes share a history of specialized suspensory arboreal locomotion. The indications are clear that the ancestral middle Miocene hominoids were structurally and functionally advanced in this locomotor style; the currently popular view that they were comparable in limb structure to the living semibrachiating monkeys is no longer tenable." Another current view, that terrestrial knuckle-walking in the manner of the African great apes formed a part of the conjoint ancestry of man and apes (Washburn, 1967, 1968) is thus also doubtful.

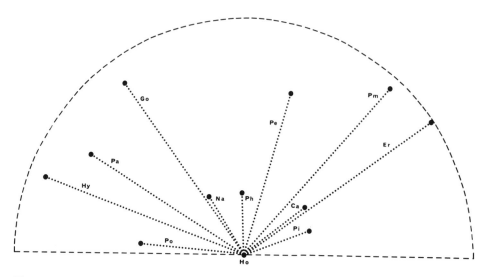

Figure 100. Relationship of extant monkeys and apes to man: canonical analysis of scapular dimensions. The relative positions of selected genera from man within the space defined by canonical axes 1 and 2.

Terrestrial genera include the following: Go = *Gorilla*, Pa = *Pan*, Pe = *Presbytis entellus*, Pm = *Papio*, Er = *Erythrocebus*. These are all far distant from Ho = *Homo*.

Arboreal genera include: Po = *Pongo*, Na = *Nasalis*, Ph = *Presbytis obscurus*, Ca = *Cacajao*, Pi = *Pithecia*. These are all relatively close to man. Hy = *Hylobates*, a highly specialized arboreal form, is far distant from man.

single group, for, though some workers suggest that the two are separate genera, many others are less certain. And the coarseness of the grouping (necessarily into genera) of the extant forms renders such a comparison not unreasonable. As the fossil data are not complete, the technique just described for the pelvis, where a positive position for the fossil is defined, may not be used. However, we are able to eliminate possibilities rather than suggest a true position for the fossil. This is done (Oxnard, 1969d) by allowing the fossil to take up, in turn, for each of the missing dimensions, values equivalent to those already known for different extant forms. Such a procedure can never suggest what the missing dimensions may have been. But it very definitely can eliminate many sets of dimensions as being outside the realm of possibility in the sense that the data sets so created lie outside the canonical space normally occupied by primates. (Of course, it is always possible that a fossil may lie totally outside the range of extant forms; but this is rather unlikely when one is looking at particular fossils that are closely related to extant forms like the great apes and man).

In this case the technique has shown that whatever the remaining dimensions of the fossils may have been, they are most unlikely to have been similar to those of any living monkey other than those of the spider and woolly spider forms extant in the New World. And of the various apes it is rather unlikely that the fossil was similar to the terrestrial gorilla and chimpanzee; it is a little more like that of the gibbon and considerably more like that of the orangutan. (Indeed it is with the orangutan that the fossil is most compatible.) But the overall result of this study confirms what is obtained from the previous examination of minimum morphological distances of extant forms, that is, that the morphologically closest forms are arboreal. This study of the fossils also goes further in that it suggests that many of the arboreal monkeys may be eliminated, as models, in relation to possible evolutionary pathways (figure 101).

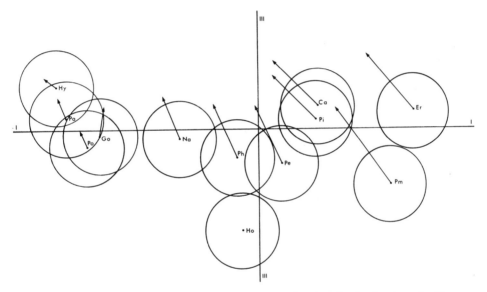

Figure 101. Effect of incorporating into the canonical analysis values for fossils: *Australopithecus* and "*Homo habilis*". Bivariate plot of canonical axes 1 and 3. Key: Po = *Pongo*, Hy = *Hylobates*, Pa = *Pan*, Go = *Gorilla*, Na = *Nasalis*, Ph = *Presbytis obscurus*, Pe = *Presbytis entellus*, Pm = *Papio*, Er = *Erythrocebus*, Ca = *Cacajao*, Pi = *Pithecia*. These particular genera represent the nearest and farthest genera of the Anthropoidea from man. The mean position for each genus is represented by the centers of the circles which are the 90-percent limits for each genus. The position of the arrow head represents the position of the combined fossil when the missing dimensions are taken to be, in turn, those of each of the extant forms. In the case of every monkey the new form falls outside the 90-percent circle and in a direction away from that of man. In the case of the apes the new form, though directed away from man, is not significantly different in this particular bivariate plot from many actual specimens of each genus. Other canonical axes narrow the correspondences even further (Oxnard, 1969).

Finally, in yet another manner, interpolation is used to examine the shoulder region. Here the different canonical axes are "relaxed" in turn and the resulting picture examined. From the matrix of extant forms, subtraction of the information available in canonical axis three, though it "moves" all of the species closer together, does not preferentially select any given species as being nearer (morphologically) to man. Likewise, subtraction of the second canonical axis provides no new data on the relationships among the groups. But subtraction of the elements of the shape of the shoulder carried by canonical axis one, while it naturally moves everything somewhat closer to man, has a remarkable and unique effect upon one genus in particular. This procedure draws the orangutan so close to man that, from its original position as about the twentieth genus away from man, it now becomes the closest form (figure 102). And the degree of this change is such that the relationships of the two forms after the manipulation are within two standard deviation units of each other. Given the extensive researches that suggest that the first canonical axis is measuring some real change that may have occurred during the evolution of the group (chapter 3), it seems inescapable that the data suggest that the easiest way to derive a shoulder like that of man is from one like that of the orangutan.

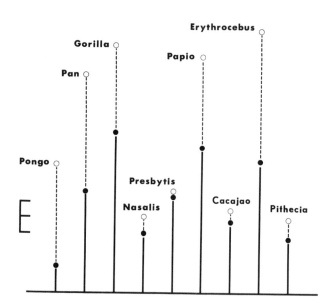

Figure 102. Effect of relaxing the first canonical variate. The open circles represent the generalized distances of the selected genera from man (the horizontal line). Terrestrial genera are distant, arboreal ones close.

The closed circles represent the change in the positions of each genus when the information in the first canonical axis is removed from the generalized distances. All genera are closer, though terrestrial genera are still more distant than are arboreal ones. *Pongo is moved preferentially to a position closest to man.* The marker represents two distance units in the vertical direction.

This does not, of course, mean that there is any actual close genetic affinity between man and the orangutan. What it very well could mean is that the presumed common ancestor of man and the African great apes is an arboreal creature capable of considerable acrobatic activity in the trees, and possessing, for that reason, a shoulder morphology more like that of the present day orangutan than of the present day gorilla or chimpanzee.

This speculation is offered because of the concordance of evolutionary suggestions when determined by all three routes: (1) relationships among extant forms (figure 100); (2) interpolation of fossil data (figure 101); and (3) manipulation of hypothetical (but now well tested) evolutionary mechanisms (figure 102). We must be clear that these procedures are less rigorous than what can be done with the fossil innominate bone. The problem with the shoulder region is not in the methods but in the specimens. Campbell (in personal communication with the author) has suggested that the scapular fragment may be even less complete than was recognized by Broom and Robinson (1950). In all of these procedures great care must be exercised in relation to the logical nature of what is being done in order to recognize and avoid circularity of reasoning.

It is of interest that in their descriptive results Broom and Robinson recognized that in a number of features of the tiny scapular fragment (e.g., the shape of the attachment of biceps muscle, shape of the coracoid process, lack of a suprascapular notch, shape of the scapular spine, and shape of lower border of the scapula near the glenoid cavity) the fossil resembles the orangutan rather than man. However, the final conclusion of these workers is that "in general structure the scapula was about intermediate between that of the orang and that of man." The above interpolative study confirms their observations rather than their conclusions.

Other Interpolative Studies: Foot Bones

Experience with these interpolative studies enables us to reassess investigations in which other workers have utilized the method apparently without being aware of some of these problems.

For instance, a multivariate study of the Olduvai Hominid 10 terminal toe phalanx (Day, 1967) may have been misinterpreted. In that study canonical analysis of several groups of apes and men demonstrates a convincing separation between them of the order of approximately six standard deviation units. Equivalent data from the Olduvai Hominid 10 terminal toe phalanx has been interpolated into that study and, in terms of canonical axes 1 and 2, appears to fall directly within the groups of men and well away from the group of apes (chimpanzees and gorillas) that were examined (figure 103). The general conclusion of that study, therefore, is that this fossil toe bone resembles those of "plantigrade men" rather than "pronograde quadrupedal apes" (Day, 1967).

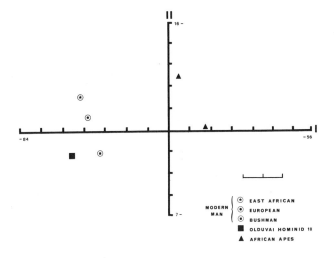

Figure 103. Canonical analysis of data from toe bones. Olduvai Hominid 10 interpolated. Bivariate plot of axes 1 and 2. Marker equals two standard deviation units.

After Day (1967); scales on axes are those of Day; Olduvai Hominid 10 falls very close to all groups of men, and this is the basis of Day's interpretation.

But the generalized distances published in that same investigation show that although in truth the fossil is nearer to the groups of men than it is to the apes, it is still far from man. Thus Hominid 10 is only as close as 26.5 D^2 units to modern man, as represented by a small Bushman sample; it is over 60 D^2 units from the nearest ape. The importance of this scale is shown by the fact that, at their nearest, both of the apes are as close to the groups of men as is the fossil (the gorilla lies within 21.9 D^2 units of the sample of fifty Europeans, while the fossil is 29 D^2 units from that group; the chimpanzee lies within 27.3 D^2 units of the sample of twelve Bushmen, less than a single D^2 unit further distant than the fossil). In other words, although the fossil is considerably nearer to modern man than it is to the apes, it is roughly as far from modern man as are the apes; there is presumably a triangular or hypertriangular relationship here that is well displayed by the model shown in figure 104. Although this reinterpretation of Day's study does not show what the relationship is between the three forms (this may be determined by including the fossil in a new analysis), it does unequivocally demonstrate that the fossil is not similar to modern man, that it is uniquely different from both of the extant groups that Day chose for comparison (Oxnard, 1972b).

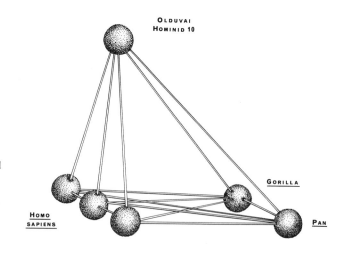

Figure 104. Model of generalized distance relationships pertaining to figure 103. The projection of Hominid 10 upon the base of the model provides a picture almost identical to that in the canonical analysis. The newly constructed model shows, however, that the fossil is not at all closely related to modern man. The generalized distance relationship exists in a space of higher dimensionality than the three shown in the model. The information that is conveyed in the higher dimensions is so small that a three-dimensional representation is possible (Oxnard, 1972). The general scale of the model is 5–6 distance units.

An interpolative multivariate study on the primate talus is also available. In this case canonical analysis is performed upon groups of men and apes, and comparison attempted by interpolating, first, data from Neandertal tali, as represented at Spy and Skhūl, and from australopithecine tali, found at Kromdraai and Olduvai[4] (Day and Wood, 1968); and, secondly, *Proconsul* tali, discovered at

---

4. The specimen from Kromdraai is referred to *Australopithecus* by most authors, but the specimen

Rusinga (2 specimens) and Songhor (Day and Wood, 1969). Once more, canonical analysis produces a satisfactory separation of man from apes. Interpolation of the fossils into the canonical analysis suggests that Spy and Skhūl are not dissimilar from modern man, that Rusinga and Songhor are closely related in shape to modern apes (chimpanzee and gorilla), and that Kromdraai and Olduvai are in a unique part of the canonical space, different from both man and apes but slightly nearer to the apes (figure 105). This result has been interpreted in biological fashion by Day and Wood who suggest that *Proconsul* tali fall within the group of "pronograde quadrupedal apes," that the Neandertal tali fall within the group of "striding bipedal men," and that australopithecine tali fall away from both of the extant groups.[5]

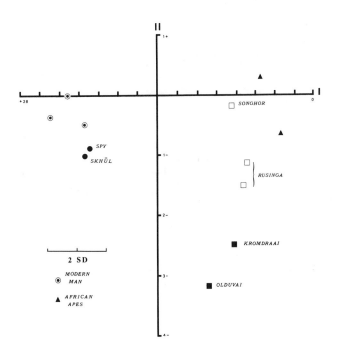

Figure 105. Canonical analysis of data from a foot bone, the talus: several fossils interpolated. Bivariate plot of axes 1 and 2. Marker equals two standard deviation units.

After Day and Wood (1968, 1969); scales on axes are those of Day and Wood, Neandertalers (Spy and Skhūl) fall very close to some modern men, *Proconsul* specimens (Songhor and two Rusinga) fall close to extant apes, australopithecine tali (Kromdraai and Olduvai) are intermediate but offset; these data are the basis of Day and Wood's interpretation.

Again, however, the generalized distances published with those same studies do not meld with the canonical analysis. It is possible to display these distances by means of a three-dimensional model which shows very clearly (figure 106) that

---

from Olduvai is, like the assemblage of hand bones, referred to *"Homo habilis"* by Leakey, Tobias, and Napier (1964). As there is some controversy about the reality of *"Homo habilis"* as an entity, the term is not used here.

5. Although Day and Wood recognize their results by describing the Kromdraai and Olduvai specimens as being neither like "striding bipedal men" or like "pronograde quadrupedal apes," their sympathies are clearly towards an interpretation that would group Kromdraai and Olduvai with men rather than apes. For although they stipulate that the unique striding gait of *Homo sapiens* had not yet been achieved, they write that the fossil talus belongs "unquestionably [to] the foot of a biped."

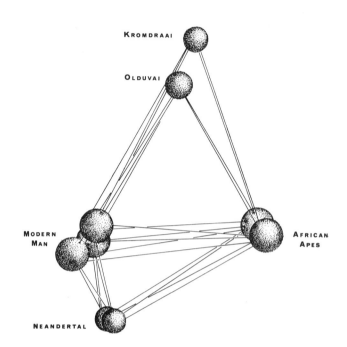

Figure 106. Model of generalized distance relationships pertaining to figure 105, with Neandertal and australopithecine specimens interpolated. This shows that the Neandertal specimens are considerably further from modern men (4+ units) and that the autralopithecine specimens are so far from both modern men and extant apes (7–10 units) that they are uniquely different from both. Again this generalized distance relationship really exists in a space of higher dimensionality than three. The information that is conveyed in the higher dimensions is so small that a three-dimensional representation is possible (Oxnard, 1972).

both Spy and Skhūl, though certainly nearer to man than to ape, are nevertheless a considerably greater distance from man (approximately four standard deviation units) than is suggested by the first three canonical axes alone (where both appear to fall within two standard deviations of the positions of the human means).

Further, this figure also shows that Kromdraai and Olduvai fall together and are a great deal further from both man and those apes (gorilla and chimpanzee) utilized by Day and Wood in this study than is suggested by the canonical axes one, two, and three alone (Oxnard, 1972).

In the case of the Rusinga and Songhor Specimens, Day and Wood do not give enough information about the D^2 values for us to be able to pinpoint the position of these fossils. But they provide enough evidence to show that these fossils lie at the intersection of a circle (obtained from D^2 values) with a plane (obtained from canonical values). Accordingly, we are only able to suggest that these three fossils could lie close together at one of two positions in relation to the groups of men and apes (figure 107, compare with figure 105). Though this study cannot demonstrate the true relationships among the extant groups and the fossils, it does convincingly show that Songhor and Rusinga are totally different from both man and African ape (Oxnard, 1972b). One of the positions obtained for the *Proconsul* specimens suggests that there may be similarities between these and the australopithecine specimens (figure 108), but, as was indicated above, this is rather a more tentative suggestion based upon the preference for one position over the other possible position for the *Pronconsul* specimens.

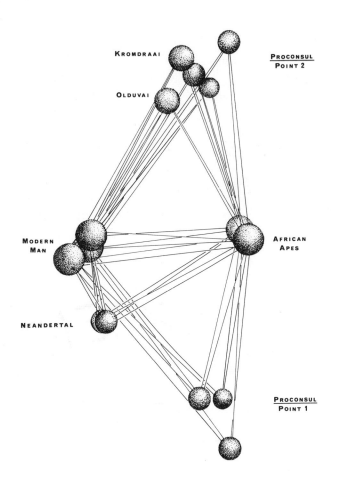

Figure 107. Another model, similar to figure 106 but with additional fossils (*Proconsul* specimens— Songhor and two Rusinga) included. This shows that *Proconsul* specimens are as far away from modern apes as are those from Kromdraai and Olduvai. As explained in the text, two positions are possible for these specimens. However, one of the two positions places the *Proconsul* specimens rather close to the australopithecine specimens. This might be of especial importance. The general scale of this diagram is the same as that of figure 106 (Oxnard, 1972).

These studies certainly deny the interpretations placed upon the analyses by these workers.

Neighborhood Limited Classification

There are thus a number of ways in which unknown data can be entered into multivariate statistical analyses of data relating to the shape of bones. In a like manner, fossil data can be interpolated into the various clustering techniques, such as neighborhood limited classification. Additional interesting results may obtain because of the fact that the data for neighborhood limited classification can be in any form or units. For instance, if radiocarbon dates are available, it is possible to allow time to be one of the dimensions in the grouping process.

Such a dimension related to time is obviously unidirectional (in this way differing from most other dimensions). This makes it possible for the technique to disconnect at a later period some data points that are ordinarily connected when

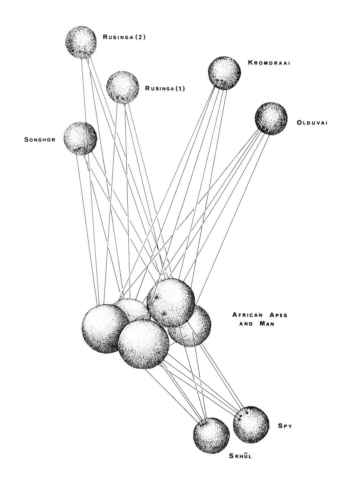

Figure 108. This is another view of the model, seen in such a way that the apes and men fall together. It demonstrates well the separation of all the fossil forms from the extant species; in addition it shows the relationship between the australopithecine specimens and position two of the *Proconsul* tali (position one is omitted from this picture; Oxnard, 1972).

time is not being considered. The method may also produce connection of certain points (that ordinarily are disconnected) following interpolation of morphological and temporal data from fossils. These possibilities are of considerable interest because they may indicate more clearly than at present such processes as convergence and parallelism. However, such analyses have not been performed; indeed, we may have a considerable wait before sufficient material of known date is available for the method to be a practicable possibility in human evolutionary studies.

Experimental Stress Analysis: the Olduvai Hand

Unknown data can also be fitted into experimental stress analyses studies. A preliminary study has been performed using constructions based upon photographs of fossils. The pilot work, insofar as it has gone, gives results which are fairly unequivocal. The following is a brief description of the method.

Into the previous stress analysis studies of phalanges described above (chapter

6), additional shapes are analyzed where, for each animal in turn and for each function in turn, the proximal phalanx is removed and the outline of the equivalent member of the Olduvai hand inserted. When the Olduvai proximal phalanx is entered into the chimpanzee hand, then the resulting shape—previously efficient within a knuckle-walking context—becomes markedly inefficient. Within the hanging-climbing context, however, the new shape (the fossil interpolated into the chimpanzee hand) remains reasonably efficient. When the Olduvai proximal phalanx is placed in the orangutan digital ray, the new shape created remains very efficient within the hanging-climbing context; it is markedly inefficient within the knuckle-walking context (figures 109 and 110).

An analysis of this type cannot tell us what the Olduvai phalanx was actually used for in that animal. But it can and does suggest most strongly that it is unlikely to have been used in knuckle-walking and that it may well have been used in the hanging-climbing situation. And though we cannot tell whether this applies to what that fossil had been doing in its own evolutionary past (rather than when it was itself alive millions of years ago), this is nevertheless useful information in the study of the evolution of the free forelimb of man.

Let it be emphasized that this study is entirely pilot in nature, and until examination of the actual fossils can be performed, it is perhaps even less than suggestive. *It does clearly demonstrate, however, the strategy that we may employ when the original can be examined, and accurate casts sectioned and modeled.*

A Speculative Conclusion

Finally, of especial interest in the interpolation of these materials is the consideration of the fitting together of the independent results obtained from the different investigations, for it now appears as though consistent information is being provided.

Data from the innominate bone suggest that considerable changes towards bipedalism, especially in the joint configuration, have already occurred in *Australopithecus*. At the same time, however, the analysis implies that this process was incomplete at the time of the Sterkfontein innominate bone and that this fragment still shows considerable evidence of having arisen from the free and mobile hip of an apelike ancestor; the fossil is uniquely different both from extant apes and modern man.

A similar speculation is now suggested by the modification of the interpolative data of Day and Wood (1968, 1969) on the talus. The noncorrespondence of the canonical variates with generalized distances for the fossils (examined earlier this chapter) shows not only that *Australopithecus* is uniquely different from both modern man and extant apes, but also that specimens referable to *Proconsul* are far less like extant African apes than had previously been thought. These findings are even more apparent when one considers inconsistencies between the original data of Day and Wood (1968) and those of Lisowski (1967). In other words, the talus, like the pelvis, suggests a curiously unique position for the fossil.

Even the analysis (Day, 1967) of the Hominid 10 terminal toe phalanx suggests

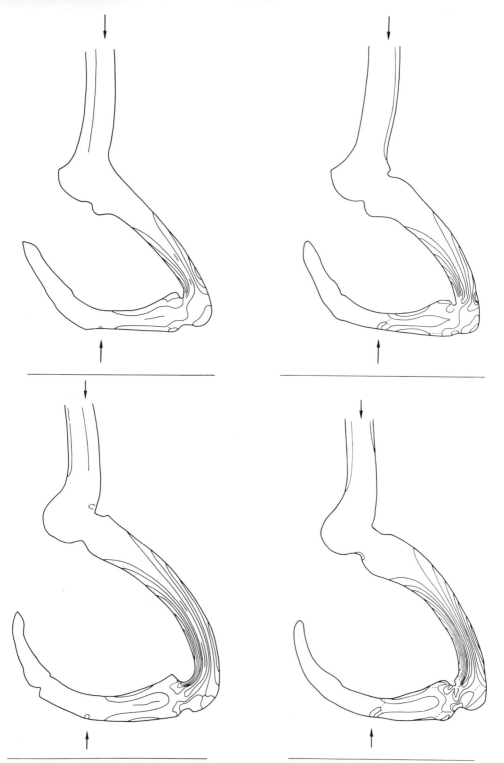

Figure 109. Photoelastic analysis of knuckle-walking simulation with fossil interpolated. Isochromatic fringes in knuckle-walking simulations of the following: upper left, chimpanzee; upper right, chimpanzee with Olduvai proximal phalanx interpolated; lower left, orangutan; lower right, orangutan; with Olduvai proximal phalanx interpolated.

In the case of the orangutan the shape is relatively inefficient (spread of 0 to 6 fringes across the model) and becomes even more so when the fossil is interpolated (0 to 8 fringes across the model). For the chimpanzee the shape is relatively efficient (0 to 3 fringes across the model); again, however, interpolation of the fossil decreases efficiency (0 to 6 fringes across the model).

Similar information appears whether the mechanical model is simple or more complex (see figure 76, chapter 6).

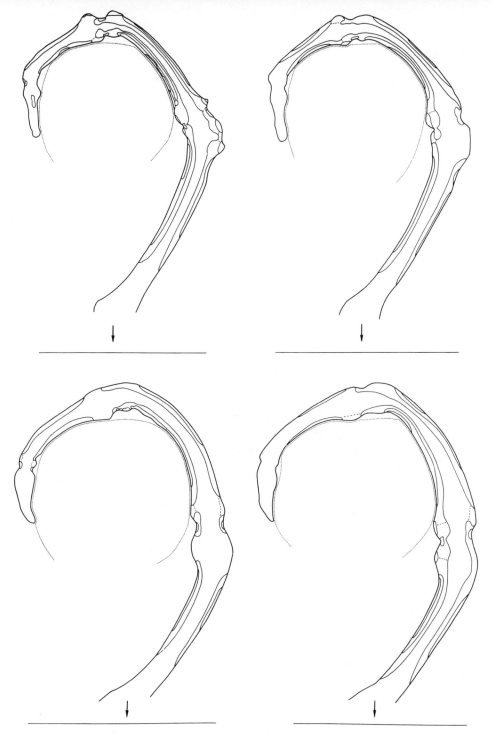

Figure 110. Photoelastic analysis of hanging-climbing simulation with fossil interpolated. Isochromatic fringes in arm-hanging simulations of the following: upper left, chimpanzee; upper right, chimpanzee with Olduvai proximal phalanx interpolated; lower left, orangutan; lower right, orangutan with Olduvai proximal phalanx interpolated.

In the case of the orangutan the shape is relatively efficient (a spread of 0 to 2 fringes across the model), and it remains relatively efficient when the fossil is interpolated (a third fringe appearing only near the neck of the bone). In the case of the chimpanzee the shape is relatively inefficient (0 to 4 fringes across the model), although it is not so inefficient for this posture as the orangutan shape is for knuckle-walking (figure 109); interpolation of the fossil in this case increases efficiency (a third fringe being again confined to the neck of the bone).

Similar information appears with both simpler and more complex models, especially those in which the existence of tendons and joints improves the mechanical analogy.

that the commonly accepted view that this specimen is similar to man rather than to apes is incorrect; once again, the correct view appears to be that it is uniquely different from both modern man and extant African apes.

Conclusions about the uniqueness of some of these fossils have also been presented by Preuschoft (1971) who utilizes theoretical stress analytic methods. Thus he reports that the form of the toe phalanx found at Sterkfontein *(Australopithecus)* suggests the possible presence of a windlass function of the plantar aponeurosis somewhat similar to what is known in man. At the same time, however, he notes that the form of the lower end of the femur of *Australopithecus* (also Sterkfontein) is about halfway between that of pongids and of man, and he concludes that we thus have "some evidence for incomplete adaptation to bipedal or incompletely abandoned quadrupedal locomotion."

In a similar manner, it is rather impressive that the independent examinations of the shoulder and hand in this work (one by multivariate techniques, the other by photoelastic analysis) also supply information that meld well with this picture. Both studies suggest that whatever the forelimb of *Australopithecus* (including the *"Homo habilis"* specimen) was capable of doing, it had not been doing it long enough for evidence of its previous functions to have disappeared. Thus both of these regions present clear evidence of functions that differ from those of modern man and that fit with recent degrees of arboreal ancestry, if not actual arboreal activity during life. Although there is nothing to suggest that these could *not* have been the shoulder and hand of recently erect but incompletely bipedal unique animals (compare with the data on the pelvis and talus), nevertheless both give information supporting ideas of function within arboreal environments that presumably relate to arboreal ape ancestry.[6]

We may thus be a considerable way toward discarding ideas such as that in *Australopithecus* bipedal *human* walking is established, or that *"Homo habilis"* is unequivocally a *human* toolmaker. Rather, we may now be able to search for the actual nature of morphologies and functions relating to a species that is becoming somewhat similar to man, but that is clearly not yet there. Recognition of such possibilities may also allow alternative hypotheses to be explored; for instance, it may yet be shown that *Australopithecus*, though a close ancestor, may nevertheless have been situated on a side branch. Recent discoveries that push human and prehuman remains back in time may eventually force this particular conclusion upon us.

---

6. Recent studies on the primate wrist (Lewis, 1971, 1972) point to similar conclusions. They negate ideas positing a degree of knuckle-walking in human ancestry or involving comparisons with semi-brachiating monkeys.

8 Optical Data Analysis in Morphology

Introduction

Although the classical method of assessing differences in morphology between one bone and another relies upon visual assessment together with extensive experience, previous chapters have shown how measurement combined with a variety of analytical tools may help to reveal unsuspected information hidden in the complex shapes of skeletal elements. It would seem that if we could depart entirely from the necessity to define reference points upon our bones for the purposes of measurements, then even more information about form and pattern might stand uncovered. One preliminary attempt that we are currently investigating is the use of Moiré fringe analysis for contouring and comparing complex shapes. That study has not yet gone far enough in our hands for its value to be clear, although in theory, at any rate, the complicated form of an object like the pelvic girdle, or the complex curvature of a surface such as that of a tooth, could certainly be characterized by this method.

Another approach that may prove useful is the method of medial axis transformation that has been used in pattern recognition studies (Blum, 1962, 1967; Philbrick, 1968). Our researches are at a very early stage, but the general nature of the approach can be visually realized from examples utilizing the innominate bone (Oxnard, 1972c). Using the description supplied in chapter 1, we can characterize a two-dimensional form through its medial axis transform. This is achieved by allowing the shape to collapse into itself in a direction normal to its boundary and at a constant rate. Such a procedure defines a "medial axis" where the opposite collapsing boundaries meet one another. The mathematical function of this axis, together with the order and speed of propagation of the collapse along it, completely defines the shape. Certainly this sort of mathematical reduction is achieved without defining any special points upon the perimeter of the form, although if it be necessary to incorporate information about such points — for example, points of possible biological import — this could be done. It may provide a mechanism for further quantitative comparison of many such shapes. The problems here are not inconsiderable; it is especially necessary to develop the technique so that it can be used for three-dimensional envelopes rather than for two-dimensional surfaces.

As a visual example of what is achieved by the method, photographs of casts of innominate bones produced by the Wenner-Gren Foundation have been treated in this way.

We may consider the stages in the production of the medial axis transform. From a photograph of the cast (plate 5, first quadrant) a two-dimensional outline (plate 5, second quadrant) may be obtained. This is then allowed to collapse into itself at a constant rate, as depicted by the ever-decreasing contours shown in the third quadrant of plate 5. Finally, the medial axis transform consists of (a) a line defined by a mathematical function and related to the major axis of the form; (b) points upon the line which give the number of the particular contour representing that part of the axis, a measure relating to the overall size of the shape; and (c) the changing distribution of points which provides the velocity of propaga-

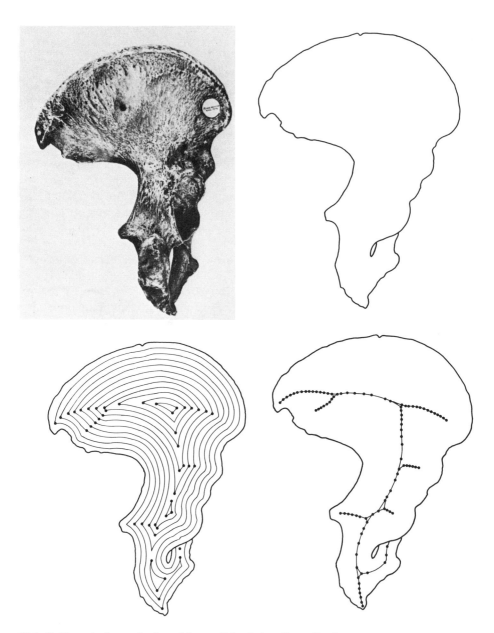

Plate 5. Stages in the production of the medial axis transform of a shape.

tion of collapse along the axis, a measure relating to differing degrees of convexity and concavity of the original outline (plate 5, fourth quadrant; see also figure 10, chapter 1).

The comparison of data once rendered in this way is a task that devolves upon further computer programs. However, to allow the reader to envisage pictorially the nature of such comparisons, the four innominate casts have been compared in figures 111, 112, and 113. The first of these shows medial axes obtained from views of the casts oriented in the plane of the iliac blade; the second is taken from the casts positioned in the plane of the ischio-pubic rami; medial axes obtained from the direct lateral view are shown in figure 113. None of these studies can as yet be used to make practical remarks about these particular bones or the Sterkfontein fossil. For instance, much more extensive studies of numerous specimens from many extant species would be required, as also would analysis of the real fossil and not of a cast. Furthermore, the information that is contained within three-dimensional replications is so important in this particular study that we must await appropriate modifications of the analytical tool. How we may cope with problems of orientation—for instance, whether we should use such arbitrary planes as that of the iliac blade (Le Gros Clark, 1964) in preference to other more objective views such as lateral, ventral, and cranial—is yet another problem. However, the series of figures illustrates the research strategy and tactic that may eventually be possible.

The Analysis of Patterns

If, however, our interest in defining form reaches into such complex realms as these, we ought also to be interested in the totality of information presented by a bone. For a bone consists of very much more than the three-dimensional envelope of its outer surface. One of the principal problems that has vexed those interested in the functional significance of bone form over many years has been the description of the architecture evident within a bone. As we have already indicated, this can be studied at many observational levels. One of particular interest is the network of trabeculae that is observable with the eye or, at most, a hand lens or low-powered dissecting microscope.

Such patterns visible in sections and radiographs of bones strongly suggest a relationship to impressed mechanical forces. Much work has been done attempting to elucidate this relationship by study of the alterations that take place in the patterns during growth and by examining the changes that can be wrought through interference in function by a wide variety of natural and artificial experiments (e.g., Murray, 1936). One of the basic inabilities of many of these studies to reach the nub of the problem has rested in the nature of the observation of the trabecular patterns themselves: how can these patterns be characterized? Usually the delineation of such patterns depends upon defining major bundles of trabeculae and the most prominent parts of the compact shell (e.g., Wagstaffe, 1874). Recently there have been utilized such methods as densitometry and radiographic scanning, but in studies aimed at the elucidation of pattern such techniques may hide the detailed answer. The primary problem, a more complete characterization of the pattern, has scarcely been tackled.

And we can well see why. For instance, one way to characterize such patterns is to measure the length, width, orientation, and position of each trabecula in a given section of a bone. These data may then be compared with those from a

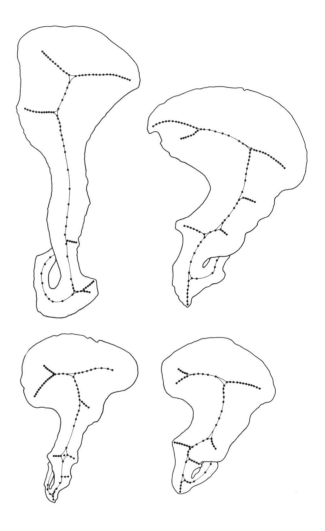

Figure 111. Medial axis transforms of the innominate bone when viewed in the plane of the iliac blade. Top left, chimpanzee; top right, modern man; bottom left, Sterkfontein innominate; bottom right, pygmy. Based upon the Wenner-Gren Foundation casts. In this view the fossil most resembles the human specimens.

number of other sections: (a) from the same bone, to obtain some idea of the three-dimensional relationships of the trabeculae; (b) from other animals of the same kind, to obtain information about the variation within species; and (c) across different groups of animals, to obtain the comparative perspective relating to the varying functions of the network of trabeculae within the different life habits of the animal species. Unfortunately, the mere problem of making such a set of measurements would be almost a lifetime's work. The difficulties of analyzing the data for even one section is a task requiring the assistance of a computer; the comparison of many such sections may take a great deal of time and effort even with a large computer. It is true that stereological formulae (Underwood, 1970) are available that may help in the delineation of the three-dimensional situation by analysis of measurements on parallel or random sections; but even here most

Figure 112. Medial axis transforms of the innominate bone when viewed in the plane of the ischio-pubic rami. Top left, chimpanzee; top right, modern man; bottom left, Sterkfontein innominate; bottom right, pygmy. Based upon the Wenner-Gren Foundation casts. Here the fossil displays resemblances to the chimpanzee.

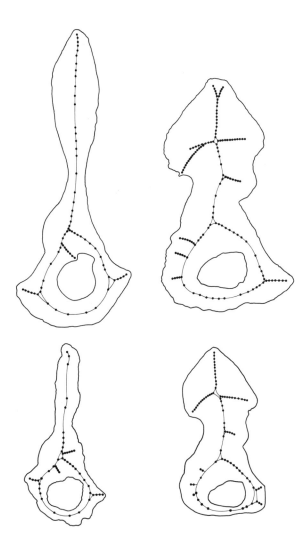

methods have not yet reached a degree of sophistication sufficient for this particular problem. In fact, this very approach to the problem is probably limited.

Other ways of analyzing data of this type include a return to the data in pictorial form. Thus a picture of a trabecular pattern holds within itself a great deal of information (much more, for instance, than is contained within any practicable set of measurements of the pattern). Very recently, numbers of investigations have been reported (many of them as collaborative studies between radiologists and computer scientists) which attempt to reveal all, especially small, details in bone sections, radiographs or autoradiographs. Often these studies have been aimed at discerning incipient pathology, for instance, early localized lesions such as metastases, or the onset of generalized bone conditions like osteoporosis. One

Figure 113. Medial axis transforms of the innominate bone when viewed laterally. Top left, chimpanzee; top right, modern man; bottom left, Sterkfontein innominate; bottom right, pygmy. Based upon the Wenner-Gren Foundation casts. Other views are also available for analysis in this way, and the resulting conclusion might well be that the fossil is uniquely different from both the ape and modern man.

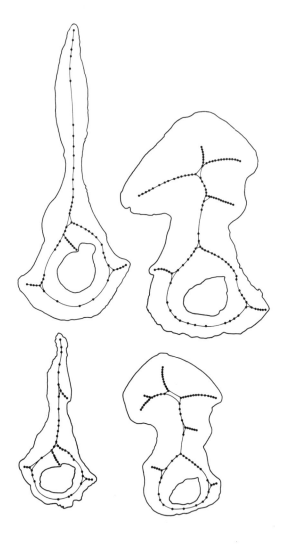

method of finding this information is to digitize the data in the section or x-ray by scanning the picture using a photoelectric cell, or by utilizing the scans formed in television images of pictures. This renders the data in numerical form and it can then be transformed utilizing computer methods. For instance, noise or other unwanted elements may be removed by means of matrix manipulations, and following such a procedure the picture may be reconstituted to reveal new information (e.g., Nathan and Selzer, 1968). Another way of doing this utilizes optical techniques whereby noise and other features may be removed by the use of physical filters; again, the aim is reconstruction of a sharpened image that provides an opportunity to visualize structures previously scarcely detectable (Shulman, 1970). Becker, Meyers, and Nice (1969) have utilized such optical filtering methods to reveal small lesions hidden within complex radiographic images of bone.

We must also acknowledge that nonpictorial methods for solving this type of problem are also being developed, such as microdensitometry and radiographic scanning. These are being employed to render more amenable to investigation such disease processes as osteosarcoma (e.g., Butler, 1968).

Most of these techniques have been pioneered in other fields and are undergoing secondary development within biology. Both the computational and optical methods for reconstruction of sharpened images have evolved as fallout from modern technological advances related to such things as the exploration of space, and from the development of instruments such as lasers and computers. The best-known examples of this are found in the transmission and improvement of pictures taken by artificial satellites and space probes (e.g., Andrews and Pratt, 1969; Andrews, 1970 a,b), in pattern recognition studies utilizing powerful computers (Rosenfeld, 1969), and in holographic and photographic investigations using optical data processing (Shulman, 1970).

A by-product of many of these methods, however, is the realization that in the procedure of reconstruction an intermediate stage exists in which the pictorial data is transformed into a nonpictorial form. Both computational and optical filtering, for instance, are applied only after the original picture has been transformed in some quantitative manner. It is this intermediate quantitation that provides an analysis of a picture. In the case of bones, this intermediate transformation may supply succinct yet comprehensive information about the details contained within complex trabecular lattices.

These analytical applications of the methods have been started in fields other than anatomy and anthropology. Some of the advances are in the nature of direct analyses of the data in a picture. The general purpose programs of Shelman and Hodges (1970) are capable of providing directly such information as relates to surfaces (e.g., area, perimeter, and moments) or to shapes (vectorial representations of shape). Other studies utilize the known properties of particular mathematical transformations to reveal the information in a picture in a particular manner. Thus computational transforms (such as Walsh functions) have been used to analyze the complex patterns in sound spectrograms (Campanella and Robinson, 1970, 1971), and optical techniques (optical Fourier analysis) have been used to study detailed patterns in different rock formations (e.g., Preston, Green, and Davis, 1969; Davis 1970). Within a relatively short period, a considerable volume of literature has sprung up relating to the analytical uses of these transformations (e.g., see recent symposium volumes edited by Thomas and Sellers, 1969; Tolles, 1969; and Yau and Garnett, 1970). It is clear that these methods are also applicable to biological data such as in the examination of trabecular structures within bones.

The method of optical data analysis is capable of the detailed data collection necessary for such studies; in addition the technique supplies its own analysis. Both data collection and analysis are performed in a manner that is almost instantaneous, and quickly and easily understood comparisons of one data set with another may be provided. This use of optical data analysis within anatomy is as yet at an early stage, but the clarity of the results and its very obvious value in a whole host of problems relating to the comparison of form of primates within an evolutionary context are so great that a description is worth attempting at this time.

Optical Data Analysis: the Basic Technique

Let us suppose that we have a complex picture of some patterned object that we wish to analyze. Let us suppose further that the picture comprises a number of small black ellipses fairly closely packed so that there is a series of clear white channels between them.

Imagine next that the orientation and position of the ellipses are random, but that the sizes of the ellipses are normally distributed around a mean value. This information may be revealed in various ways: (a) by making a large number of measurements of the x and y coordinates of the positions of a sufficiently large number of ellipses to characterize the pattern (analyses of these data would show the random nature of the positions); (b) by making a series of angular measurements of the directions in which the ellipses are pointing with reference to the x and y coordinates (here analysis could suggest the random nature of the directions); and (c) by measuring a sufficiently large number of ellipses (appropriate analysis could demonstrate the normal nature of the distribution of their dimensions). The three sets of data, each in itself a considerable undertaking to acquire and analyze, would all be necessary to our understanding of the pattern.

An analysis equivalent to the one outlined above can be performed by optical data analytic methods utilizing some of the properties of convex lenses and coherent (laser) light. The core of the technique is this: the many details of any black and white transparency can be transformed into their visual Fourier equivalent by appropriate treatment with an optical system and coherent light. This is a mode of analysis which would not immediately be obvious to one whose analytical background has been primarily within the realm of Gaussian statistics.

Let us suppose initially that we merely wish to analyze a single one-dimensional scan of this complex two-dimensional pattern of ellipses. Such a scan resembles somewhat the wave pattern in a ray of light, the peaks representing the bright regions of the pattern, the troughs the dark. With certain simplifying assumptions this wave form may be analyzed on a digital computer by a Fourier transformation:

$$f(x') = \left|\int f(x) \exp^{-j\omega x} dx\right|^2.$$

In this equation $f(x)$ is the input signal in the form of a one-dimensional traverse (wave), ω is a cosine vector, and j is a complex constant; $f(x')$ is, of course, the required Fourier answer.

However, rather than carry out such an operation by computer, which necessitates digitizing the information in the wave and then performing the above procedure computationally on that data, the whole process can be performed optically; this is in fact what occurs if one passes a complex wave of light (containing a number of different primary wavelengths) through a prism to separate out the constituent simple wavelengths. In this example the prism acts as a frequency analyzer and transforms the wave forms in the ray from the time domain to the frequency domain. The diagram (figure 114) shows the analogy that is being suggested; the contribution of each wavelength to the total pattern and intensity of the complex entering wave (this is the information hidden in the complex wave) is shown by the separate intensities of each color (wave length) in the resulting spectrum (it is called an intensity, energy, or power spectrum of the light).

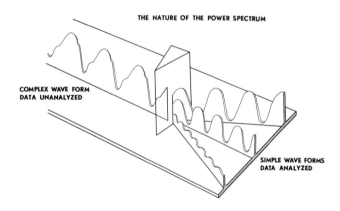

Figure 114. The analogy between the spectral analysis of white light into its component colors and that of one-dimensional optical data analysis.

Because a single traverse across the black and white picture resembles the complex wave of light in the analogy, it can be analyzed in a similar manner, provided a suitable analogue to the prism can be found. In this case we are transforming from a one-dimensional spatial domain to a frequency domain.

Of course, the real data (a picture of black particles and white pores) which we are interested in analyzing does not exist as a one-dimensional configuration. It is the full body of a two-dimensional array where the data consist of x and y coordinates (together with information as to the blackness or whiteness of the picture at each point). Using the same conventions as before, the power spectrum of a two-dimensional signal can also be expressed by a Fourier equation:

$$f(x',y') = |\iint f(x,y) \exp^{-j[\omega x + \omega y]} dx \cdot dy|^2.$$

Here the input signal $f(x,y)$ is the continuous picture. If the input signal is discrete, this equation also may be solved by applying finite Fourier methods to a large metrical data set on a digital computer; $f(x',y')$ is the appropriate Fourier answer in this case.

However, it is not necessary to rely upon mathematical manipulation to create Fourier transforms of such data. One property of any lens forming a real image is that it performs a Fourier transform on the input signal.[1] The geometric arrangement of lenses necessary to produce exact Fourier transforms is shown in the diagram (figure 115). The transform (power spectrum) may be recorded as a pattern of light on photographic film.

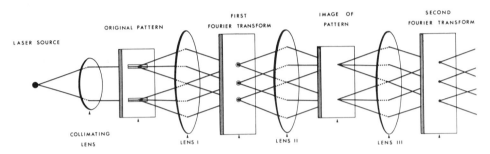

Figure 115. The optical arrangement used in two-dimensional optical data analysis.

Using such an instrument, specified visual items in the original picture that are defined by size or orientation can also be identified in the transform. This identification can be quantified so that the contributions of the specified items relative to those of the whole picture may be obtained. Thus the original picture is represented by the total flux (or intensity) of light across the entire area of the transform; that is, the brightness of the transform may be thought of as a surface defined by a z axis representing degree of brightness and erected upon the x and ω axes (polar coordinates) representing the area of the transform. The flux or the intensity of light representing the original picture is then given by the volume lying between the surface and the transform plane. For many images a photograph of the transform is a circle of light with a bright center; the intensity, therefore, is the volume under the bell-shaped surface that can be erected over the circle. Clearly, smaller volumes may be defined over those parts of the transform representing specified visual items only. The relative contribution of the specified items to the whole picture is given by the ratio of the appropriate volumes.

In terms of our example utilizing the picture of the ellipses, we may wish to know the relative contributions of ellipses between two size limits. The size of particles within the original picture is given, in a transformed manner, by distance from the reference axis of the transform (x of the polar coordinates). Accordingly, the contribution of ellipses between the size limits in the original picture is represented by the volume above a circular band on the transform representing the appropriate size limits (figure 116). It is thus simple to distinguish between transforms representing pore and particle conglomerates having the same porosity but

1. The complete optical Fourier transform is a function consisting of both amplitude and phase spectra. Phase relationships are extremely difficult to analyze. Fortunately, however, the amplitude spectrum can be mapped onto photographic film.

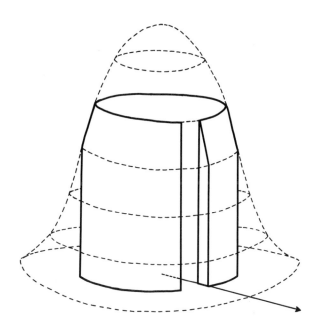

Figure 116. Picture size and the Fourier transform. The original picture is represented by the flux of light passing the transform plane, that is, by the volume of the entire bell-shaped body erected upon the circular base. Elements between certain sizes are represented by the volume contained within the solidly outlined object defined by the appropriate contours on the bell relating to particle size.

different pore-size contributions. It is also possible to quantitatively identify and measure the source of such a difference.

In the same manner, directional or vectorial properties are preserved in the transform. In this case the relative contribution of all ellipses, when measured between certain angles within the original picture, as compared with the entire picture, can be determined. This contribution is given by the ratio of the volume of the power spectrum above a sector-shaped region (representing the original angular limits, again appropriately transformed) as compared with the entire volume above the transform (figure 117). This property thus allows quantitative assessment of directional nonhomogeneities within the original pattern.

Obviously, then, the relative contribution of particles which are of specified size and which are oriented at specific angles can be obtained from the volume above the appropriate segment (direction) taken from the appropriate circular band (size) on the transform. The application of an analysis of this type to photographs of complex biological patterns may be most revealing; it is in this way that size and orientational information relating to pattern may be uncovered. An excellent synopsis of Fourier optics is provided by Goodman (1968).

Finally, there is a question as to how the information in the transform, which is visually very obvious, can actually be translated into quantitative terms. This is done by calibrating the method. First, it is necessary to identify x and ω coordinates of the circular area of the power spectrum. This is given by the power spectrum of ruled gratings of known spacing (figure 118). Such calibration shows (a) that the distance between the gratings is represented by the inverse of the distance x from the reference axis of the transform, and (b) that the angulation of

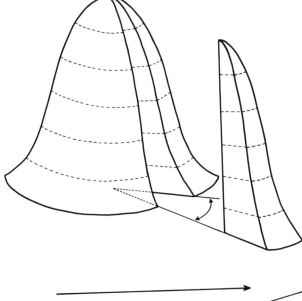

Figure 117. Picture orientation and the Fourier transform. Here the elements of a certain angulation are represented by the volume of the "slice" taken from the bell.

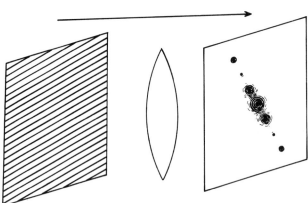

Figure 118. A diagram of a simple grid and its power spectrum (shown reversed as a black pattern upon a white background; the actual transform would be light spots upon a black background; see plate 6). The distance between the central ray and the first harmonic is related reciprocally to the distance between the grid lines; the angulation of the set of harmonics is that of the grid lines plus 90°.

the gratings is at 90° to the angle ω on the transform. In this way the coordinates of the area of the transform are defined.

The second requirement is to determine the z axis defining the surface above the transform which is formed by the flux or amount of light passing the transform plane. This may be done by passing sensitive recording photometers across the focal plane of the transform. It may also be obtained from a photograph of the transform by reading onto microdensitometers. In the current studies it is being found by using a televisionlike device. The IDECS processor of the University of Kansas is used to contour the power spectra at a number of energy levels. (This instrument, the Image Display Enhancement and Combination System, has been developed for the analysis of multispectral aerial photography and radar imagery).

Using the system, visual images in black and white can be transformed into images in color, in which case the colors are used to calibrate different levels of intensity in the original black and white picture. It is in this way that we have contoured the intensity of light across the transforms. More sensitive techniques are currently being investigated, but for biological purposes the IDECS images appear to be satisfactory. The use of this technique means that the information, rather than displaying the intensity (volume) of light for a given particle size, displays the area of particle representation for a given intensity of light. This display of the area of particle representation may prove more useful for many biological purposes; it would not, in any case, be difficult to ascertain the intensity of light for a given particle size, if necessary, in a given situation.

In the case of the theoretical example that we are considering (black ellipses, the size of which are normally distributed, but which are randomly placed on a white background), the power spectrum (the analytic result) consists of a series of light patterns so arranged that they form a thick ring of light around a central point—the reference axis. The fact that the ring is completely circular without any directional parameters demonstrates the random arrangement and directions of the ellipses within the field. The distance of the brightest center of the thick ring of points from the reference axis gives a transformed measure of the mean size of the ellipses. A measure of the dispersion of light away from the central core of the ring provides a transformed measure of the variation of the sizes of the ellipses. All this may be obtained from visual inspection. Appropriate transformation calculations and a few measurements relating to calibration supply the data in figures.

Further examples that demonstrate information obtained from pattern are shown in figures 119 and 120. Here structurally repetitive patterns (periodically regular) have been analyzed to show the relationship between the original picture and the power spectrum. In each case size and orientational information may be clearly discovered (Pincus, 1969).

Figure 119. The power spectrum of a simple pattern. Again, orientational and size information is obvious.

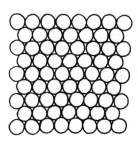
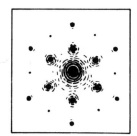

An Example from Geology

These techniques are being developed essentially in science areas other than biological morphology. In many cases the rationale behind development relates to problems of image enhancement—that is, the use of the power spectrum in the

Figure 120. The power spectrum of a pattern with distinct orientational features.

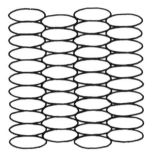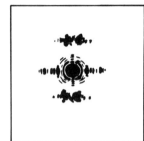

reproduction, usually in an improved form, of the original image (e.g., Shulman, 1970). However, in a number of applications within geology, the technique has been modified so as to be analytical in nature; the modifications are carried out to avoid the long, difficult, and expensive use of digital computers in the analysis of pore and particle states in thin sections of oil bearing rocks. One set of such researches is being carried out at the University of Kansas by Dr. John Davis and his associates, and it is to this group that the author owes thanks for being able to make the optical data analyses of sections of bones which are described in later pages.

The geological work itself shows clearly how valuable the technique may be. The next picture (figure 121) shows, above, the original photographs of thin sections of two rocks, and, below, the appropriate contoured power spectra for those same specimens. It is obvious to the eye that the two rocks are different; no elaborate analysis is necessary for the differentiation. However, the differences between the contours, while equally obvious, do allow quantitative assessment of the precise nature of this very considerable difference.

But the real power of the technique can be seen from the second geological example (figure 122). In this figure the two sections of rocks, above, are extremely similar—so much so that the casual observer is hard put to differentiate them, let alone say what is the precise nature of the difference. Examination of the contoured power spectra, below, confirms that they are generally similar. But it also shows that they differ in a special way. Although most contours for both specimens are circular, the most central contours for the specimen on the left (the 10 and 20 percent contours for pores about 0.5–0.25 mm in size) are elliptical with the long axis vertical. The difference, therefore, between the specimen on the left and that on the right resides essentially within particles of that size; those on the right are random in orientation and in this feature do not differ from the other particles; those on the left are oriented to a degree with their long axes tending horizontally and thus differ from the other, smaller, particles which are arranged randomly within the section as in the other section. This can be seen with hindsight, I think, when the original pictures are reexamined. It is most doubtful, however, if the result would be reached in advance of using the method, a fascinating example from geology of the sensitivity of the technique (Preston, Green, and Davis, 1969; Davis and Preston, 1970; Davis, 1970; Davis and Preston, 1971).

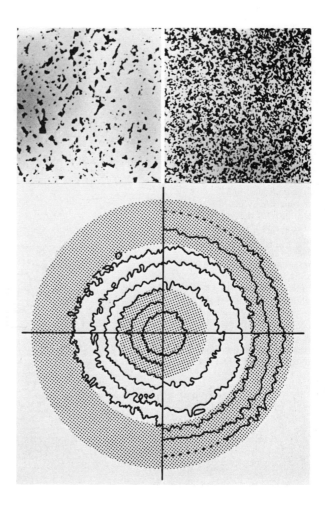

Figure 121. An example from geology. Photographs of two sections of rocks that are markedly different. The contoured power spectrum of each is shown on each respective half of the diagram below. The differences in size of the pattern elements are obvious from the differences in radius of the contours. That the contours are indeed circular demonstrates that the particulate information is essentially random. (Courtesy of J. C. Davis.)

Some Preliminary Investigations of Trabecular Pattern

At this stage in these investigations of trabecular arrangements in bone, little real work has been done. At most it has been shown that the technique will indeed work for biological data and that the nature of the results that flow from such studies can be demonstrated (Oxnard, 1972d). Two examples of the analysis of bone are so far available. On the one hand the comparison is attempted of two pictures of bone trabeculae that are rather different and where, in terms of diagnosis, there can be no mistake: the transverse sections of the third and fifth lumbar vertebrae respectively. These two sections differ considerably because the fifth lumbar vertebra intergrades morphologically and functionally into the sacrum. A comparison is also being undertaken, on the other hand, of the differences between sagittal sections of the second and fourth lumbar vertebrae. In this case the differences are minimal, and it takes an experienced anatomist to decide which is which.

Figure 122. A second example from geology. Photographs of two sections of rocks that are so nearly similar that it is difficult to tell by eye that they are different. The contoured power spectrum of each is shown on each respective half of the diagram below. The similarity in size of the pattern elements is clear from the general similarities of the contours. Small differences may be quantitatively defined from the contours if necessary. But the major differences relate to the smallest contours of the specimen on the left: these are elliptical and show that the largest particles of all are preferentially oriented, as compared with the circular, random arrangement in the section on the right. (Courtesy of J. C. Davis.)

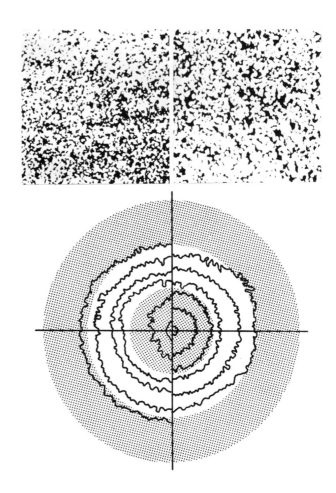

The diagram (plate 6) shows, above, transverse sections of the first two vertebrae, displayed so that the left half of the third lumbar vertebra is compared with the right half of the fifth. Below, the appropriate power spectra are shown in the same way. (It must be appreciated, however, that the power spectrum that is shown is one half of the spectrum of the whole section in each case). This mode of comparison suggests immediate differences in the size of the power spectra. Of course, this merely relates to the differing sizes of the components of the two sections; the elements (trabeculae) of the third lumbar vertebra on the left are the larger, and this is confirmed by its smaller power spectrum. Other differences are also apparent. For instance, the power spectrum of the third lumbar vertebra is more nearly circular in outline than is that of the fifth. This departure from circularity for the fifth lumbar not only is a greater trend towards an oval shape but also includes a wedge-shaped deficit lying a little above the horizontal. These features are somewhat difficult to see in the pictures as reproduced; when we view the originals, however, there is no doubt that they are present. They are rendered

Optical Data Analysis in Morphology

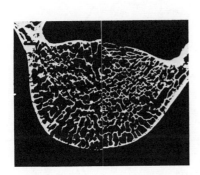

Plate 6. The power spectra of lumbar vertebrae. Comparison of transverse sections of two human vertebrae (3rd and 5th lumbar) that are markedly different. Comparison of the power spectra also shows the obvious difference.

obvious by the method of contouring the spectra as is shown in figure 123. Here the size and shape differences described above are clearly visible; they may be measured for particle size and orientation; and for any given particle size or orientation a highly detailed result may be obtained if necessary. Because this example is being used in an illustrative manner with no further hypotheses in mind, additional quantitation has not been undertaken. The description certainly confirms that the two pictures which we would not confuse on simple inspection are also not confused by the technique.

The next figure (plate 7) is a better example of the discriminating power of the method and shows the two rather similar sagittal sections of the second and fourth human lumbar vertebrae; below them are the corresponding power spectra. The only easily discernible differences between the actual sections themselves relate to the generally smaller trabeculae in the fourth lumbar vertebra as compared with the second. This information is also immediately obvious from the power spectra: the power spectrum of the fourth lumbar vertebra is larger than that of the second lumbar vertebra. The actual measurement of this difference could be carried out much more easily on the power spectra than upon the original sections themselves.

In addition, however, the spectra demonstrate interesting directional differences between the two sets of pictorial data. First, both spectra display a general departure from circularity, which reflects differences between those trabeculae which are at right angles to the craniocaudal axes of the bones as contrasted with those that are parallel. This too can be discerned in the original pictures. Trabeculae that

186   Form and Pattern in Human Evolution

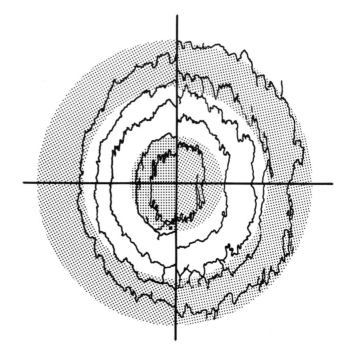

Figure 123. This diagram shows the contoured power spectra for the comparison described in plate 6. Quantitative differences between the two power spectra (left and right halves of the diagram) may easily be made.

are at right angles to the craniocaudal axis are much shorter; this is shown by the inverse elongation in the transform in the direction at right angles to them. However, in the transform of the second lumbar vertebra, the elongated region is not so long as in the fourth (i.e., trabeculae oriented at right angles are bigger). The elongated region is, in any case, slightly inclined in the second lumbar vertebra as compared with the fourth (i.e., the directions of these orthogonal trabeculae are offset in the second as compared with the fourth). Finally, a new set of orthogonal elements is present in the second lumbar vertebra; these are centrally located and are again slightly inclined (and relate to differences between the layers of compact bone in the two examples).

These components of the bone structure are obvious in the original pictures only with the hindsight of these investigations. Because they are associated with orthogonal but inclined sets of trabeculae and bony plates, they are likely to have functional relationships to the stresses borne by these two different bones. And although the reader may realize that these facets of the power spectrum cannot be too easily seen from mere inspection without some experience, the contours of the power spectra (figure 124) leave no doubt that these differences genuinely exist. The contoured power spectrum of the fourth lumbar vertebra resembles at all energy levels an American football; that of the second lumbar yields generally inclined contours at certain energy levels. Considerably more information is available from these contoured spectra, but it remains for further investigation to seek these out.

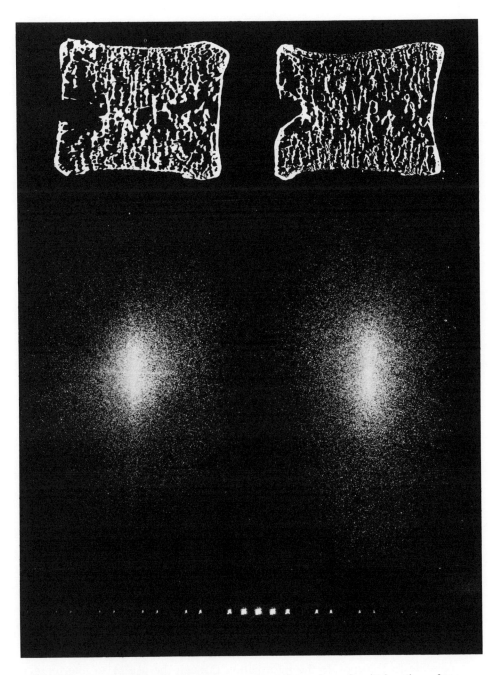

Plate 7. The power spectra of further lumbar vertebrae. Comparison of sagittal sections of two human vertebrae (2nd and 4th lumbar) that are rather similar. Comparison of the power spectra displays more clearly the differences. Below is shown the power spectrum of a 1mm grid for calibration.

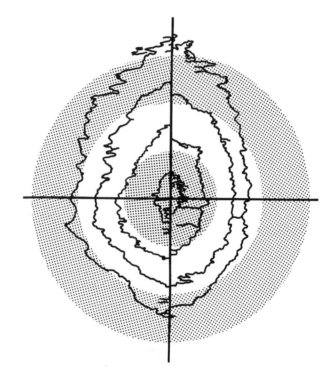

Figure 124. Comparison of the contoured power spectra of the two vertebrae shown in plate 7. Quantitative differences are now most obvious.

Although it is possible to extract particle size and orientation distribution directly from the intensity or power spectrum, problems of interpretation may result from the fact that the extreme coherence width of laser light creates spurious responses in the Fourier plane resulting from instrument, film, and image interactions. In addition the intensity spectrum is not an additive function of spatial frequencies in the original if the coherence width of the laser light extends beyond the sizes of the particles in the image. These problems produce, for instance, the "speckle" effect that is obvious in the power spectrum, especially in its background (plates 6 and 7). Both problems can be eliminated by the use of partially coherent light; the intensity spectrum which is then produced is proportional to a smoothed power spectrum of spatial frequencies in the image.

A second problem relates to the fact that although the Fourier transform at the focal plane contains all the information in the original picture, the photographic reproduction of the transform by the interpolation of a camera at that point records only the frequency components. Information that is present in the phase element of the Fourier transform is therefore lost. However, this information relates to localities within the picture, and though it cannot be examined directly, it is possible to incorporate the information through the use of an interference technique. This procedure, two-dimensional holography, is being used in pattern recognition studies (e.g., Holeman, 1968).

In both of the initial studies described here, the procedure, although analytical, is being used in an exploratory manner; that is, the technique is employed in an empirical way as a searching tool. The power spectrum is a "finger print" of the trabecular pattern for the purpose of comparison, though it is true that we are already attempting to interpret the fingerprint in terms of the structural differences between the pictures.

The technique may also be used explicitly as a hypothesis-testing mechanism. For instance, in the geological studies that have been quoted, the guiding theory is that the pore network is the realization of a random process. From the power spectrum, unique and sufficient parameters may be obtained that characterize the pore network and afford a test of the actuality of the random nature of the rock fabric. In exactly the same way, utilizing the technique for the study of structure in bone may provide possibilities for testing hypotheses as to the nature of the bony patterns. Such tests devolve upon the similarity or dissimilarity between the *observed* spectra and those that might be *expected* given particular models as to the nature of bony (trabecular) structure.

Alternative theoretical models may well be tested in this way. For instance, one obvious theoretical model worth examining is that relating to orthogonality of bone elements, although it is, at best, only an approximation. Perhaps of greater interest at this time is the testing of experimental models. In this case, rather than making theoretical assumptions about bone structure, experimental hypotheses might be put forward (possibly on the basis of procedures such as stress analysis, or again in relation to some experimentally defined model of the growth of a part) to be tested against observed power spectra.

Although the power spectrum of the sagittal section of the fourth lumbar vertebra shows only the orthogonal arrangement that we already think we see, the power spectrum of the sagittal section of the second lumbar vertebra is more complicated (figure 125). Though the smallest contours, related to the largest picture elements, demonstrate the cranio-caudal arrangement that is noticeable in the original section, and, likewise, though the largest contours, related to the smallest picture elements, show a similar orientation, there is a middle contour that suggests that certain intermediately sized elements are oriented at 32 degrees to the general orthogonal arrangement. The presence of these additional features of the second lumbar vertebra seen in sagittal section, as compared with the fourth, may well be related to differences in the degree of bending suffered by these vertebral bodies as a resultant of function in an upright posture over a period of time. Such morphological information is thus capable of testing the hypotheses generated by stress analytic analogies of various types.

These fascinating possibilities have not yet been completed. Nevertheless, with a tool of this sort they are possible. With hypotheses-testing methods added to the empirical solutions for the identification, characterization, and comparison of complex biological patterns we have powerful weapons for the morphologist.

Theoretical Applications to Problems of Bone Morphology

Data from a nonrandom arrangement such as is presented by bony bars and marrow cavities within a bone produce a rather more involved power spectrum than

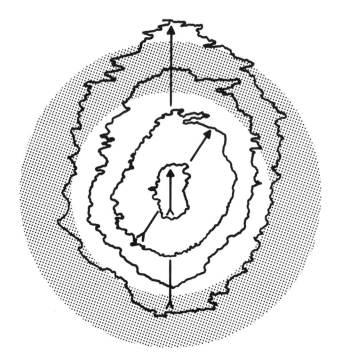

Figure 125. Contoured power spectrum of sagittal section of second lumbar vertebra.

does the theoretical pore and particle case discussed above, or the actual geological example. Bone possesses a structural regularity not present in most pore networks. This regularity is not periodic, or at least not in the sense of a diffraction grating, a microscopic organization such as a biological membrane, or a helical molecule. Nevertheless, a photograph of the power spectrum of a trabecular pattern can be analyzed far more easily for properties relating to size and orientation of trabeculae than can the original photograph of the pattern. As has been pointed out, it is a relatively simple matter to have even this analysis performed automatically, using the IDEC System to superimpose contour lines relating to the brightness and position of different elements. Such analysis can be carried out very quickly, and many sections can therefore be readily examined.

Even the need to compare many sections of one specimen can be circumvented. It is clearly possible to analyze photographs of radiographs which contain in a planar fashion the projected data for the full specimen. As an intermediary in the leap from a section to a radiograph it is also possible to examine tomograms of these structures. Tomograms are radiographs on which, by movement of both x-ray machine and film, we are able to isolate sections of the trabecular pattern. Current limits on the thickness of such sections are of the order of one millimeter, and hence they represent a genuine intermediate between a surface view of a cut specimen (of zero thickness) and the traditional radiograph of full object thickness.

But the technique can be taken very much further. It is possible to utilize filters to screen out some data, thereby allowing other information to present itself the more readily (Dobrin, 1968). For instance, if it is obvious that a certain number of the elements (trabeculae) are clearly oriented in a particular direction, a filter can be inserted which will screen out all items oriented in that way. The resulting picture may allow one to see more clearly what remains (plate 8). Again, it may be that the pattern of a certain size of trabeculae is very obvious; once more appropriate filters can be used to screen out items of that size, permitting inspection of the pattern of other smaller trabeculae (figure 126).

Such filtering techniques may be of particular importance in the detection of incipient systemic bone diseases where the resulting lesions often relate to alterations in the laying down of bone and may well produce their earliest changes within the smallest trabeculae. In bone resorption, large trabeculae may become only slightly thinner whereas the smaller may completely disappear. These changes cannot be detected by routine methods but with a technique where the contributions of the smallest trabeculae can be readily examined, early diagnosis may well be a simple matter. Incipient osteoporosis may thus be detectable in this way. A theoretical example of such filtering for removing random noise (figure 127) or of a regular pattern (plate 9) with reconstruction of originals is provided by Shulman (1970).

Filtering techniques can be taken a step further. The power spectrum of one specimen can itself be used as a filter for the examination of a second. This process then results in, first, the reconstruction of the image of the actual differences between the specimens; it is not clear at this point exactly how this latter procedure would present itself, but it certainly would allow, for instance, studies in which we might compare similar pictures from specimens of vastly different size and thus investigate the effects of change in size upon the comparisons. It may prove important to carry out such analyses as a necessary test of the methods. Secondly, the process can result in the production of the power spectrum of the differences between specimens. From this may be calculated the parameters of the differences. Such a comparison consists of the true differences of all the many elements, not just those of major trabecular bundles. This is truly comparative anatomy at the turn of a switch.

Finally, there is a very special application of filtering methods in terms of the analysis of sections or radiographs of fossils. Such pictures often delineate very clearly trabecular and other architectural patterns within fossils. This is especially the case in the remains of large dinosaurs and in bones found in the tar pits in California. But with most primate fossils, though an internal structure may be discerned, it is often obscured by crystalline and other patterns relating to the process of fossilization; analysis therefore is difficult. With a technique such as the optical data processing and filtering described above, it is possible to subtract the power spectrum of the crystalline structure from that of the entire fossil picture. What is left relates to those elements of the pattern that are associated with the original architecture of the bones. It may even be possible in some cases to examine pressure distortion effects with such a technique.

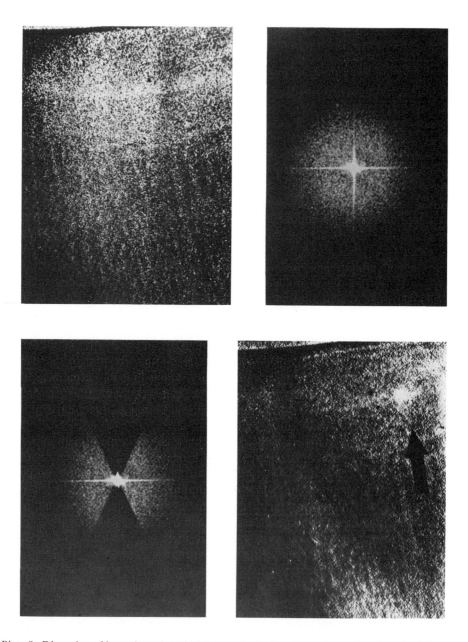

Plate 8. Dissection of bone through optical data analysis. A radiograph (much enlarged) of the following: upper left, a piece of bone; upper right, its power spectrum; lower left, the power spectrum after filtering of a sector; and lower right, the radiograph reconstructed from the filtered spectrum. A small structure (indicated by arrow) that is hidden in the initial radiograph is revealed by filtering. (Courtesy H. C. Becker, P. H. Meyers, and C. M. Nice, Jr.)

Figure 126. Dissection of a bone by computational methods. A section of a vertebra (above left), together with reconstructions in which vertical elements (above right) or horizontal elements (below) are displayed. In this case the "dissection" was carried out computationally by matrix manipulations. (Courtesy of R. H. Selzer.)

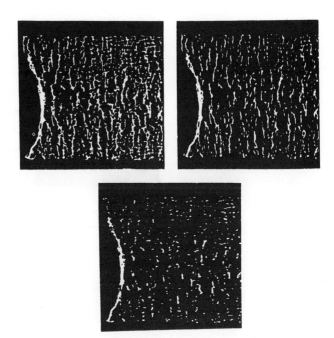

Such comparative studies may also yield two kinds of information. On the one hand, the power spectrum of the difference may be used to generate a new image: this is a picture of the trabecular patterns with the clouding factors removed; such an image may well be of great interest to those palaeontologists who wish to know more of the bone as it was. On the other hand, the power spectrum of the difference could itself prove of interest because it represents the appropriate analysis of the fossil trabecular pattern, a result of interest to the functional anatomist. We cannot yet know to what extent these hypothesized investigations may be of use. But, taken at face value, there is no obvious reason why they should not work.

These various possibilities are analogous to the way in which these techniques have been used to get rid of unwanted patterns in geology and astronomy. Dominant effects, such as vegetation, tending in one direction in aerial photographs of the earth may be removed to reveal marginal indications of geologic outcrops tending in other directions; similarly, unwanted noise in pictures of the lunar surface taken by satellite may be filtered to obtain better definition of detailed features of rocks barely visible in the original photographs (Dobrin, 1968).

Further Speculations on Analysis of Form and Pattern in Biology

Throughout this chapter I have oriented the discussion towards the analysis of relatively complex patterns. This relates, of course, to two facts: first, that I was introduced to optical data analysis by J. C. Davis through the study of pore and particle states in thin sections of oil-bearing rocks, and, secondly, that I have for

Figure 127. Removal of "noise." A complete Ronchi ruling (first), a similar ruling to which has been added a large element of noise (second), and the ruling that can be reconstructed from the "noisy" picture by appropriate filtering (third). (Courtesy of A. R. Shulman.)

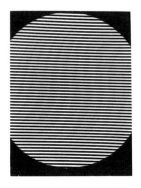

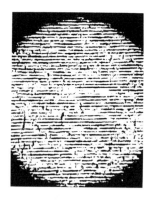

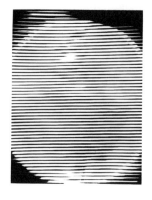

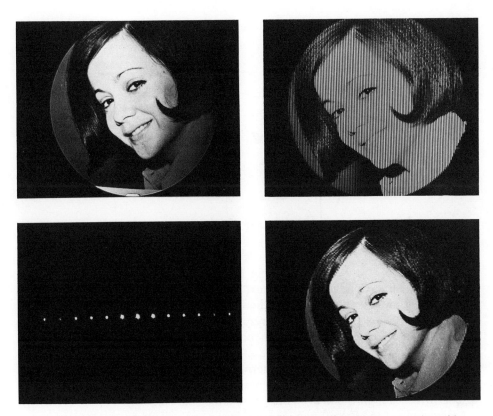

Plate 9. Removal of rulings from a picture. Television picture showing scan lines together with the removal of the lines using spatial filtering and lens rotation. (Courtesy of A. R. Shulman.)

many years had considerable interest in the investigation of somewhat similar-looking trabecular patterns in bone. There are analogies between the patterns in oil-bearing rocks and in cancellous bone. It would seem that there are many other biological patterns of complex form that may be analyzed in this way; branching patterns in trees, in corals, and in Purkinje cells spring immediately to mind.

However, it is also obvious that rather simple shapes give rise to transforms when examined in this way. And though simple shapes may be analyzed by many other techniques, some of which are described in other chapters of this book, there is no reason why this method cannot also be used. Because the method does not depend upon the observer's defining his own reference points (although it is true that he may have to define his own reference plane for the purposes of producing a photographic record upon which the method depends), the technique may achieve an objectivity unavailable to others. That this can certainly be done is evidenced by work in which homologous chromosome pairs may be identified by means of their similar transforms. The well-known example of Goodman (1968) may be quoted to show how a shape as simple as that of a 5 may be characterized;

it is not difficult to recognize the various elements of the original in the transform (figure 128).

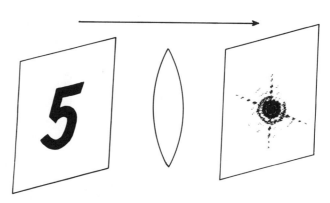

Figure 128. A figure 5 and its power spectrum (which is reversed).

In like manner, the relatively simple but irregular shapes of bones may be examined. If this is done as for the figure 5 just quoted, the only information that is supplied relates to the outline of the bone displayed as a shadow on photographic film. However, it seems entirely within the realm of possibility that a photograph of a whole bone which has been contoured (using a laser contouring technique or perhaps a Moiré fringe method) may be examined. This would certainly supply a dichotomous pattern, where the entire information would reside in the external shape and contours of the bone. It would indeed seem a logical further step in the analysis of gross biological shapes.

An example of the use of the technique in the analysis of macroscopic shape is provided by Mr. Joel D. Gunn of the University of Kansas. Outline drawings of micro blade cores from the site of Divostin, Yugoslavia, are examined.[2] The power spectra that can be obtained from such pictures are shown in plate 10, and there is little doubt that they give quantitative information about the shapes of the cores. For instance, the light secondary horizontal rays outside the main burst of light give an angle which is an estimate of the range of variation in the scar patterns upon the cores. Suffice it to say that this pilot study suggests that the method may be of interest in the examination of objects such as archaeological artefacts. Furthermore, the example suggests that whatever the nature of data, if they can be expressed in the form of a complex lattice, then they may yield to optical data analysis.

A final question that may be worth our while considering is, Why use a technique like Fourier analysis at all? Once we become enmeshed in a field like Fourier optics, two things immediately appear obvious: (a) Fourier transformations are only one of a whole series of mathematically related transformations, and (b) all of these transformations are, in principle at any rate, implementable

---

2. The site is Neolithic of the Vinca phase, and it was excavated by Dr. Alan McPherron of the University of Pittsburgh.

Plate 10. Human tools and their power spectra. Line drawings of archaeological artefacts give a power spectrum that may be of considerable value in morphological comparisons. (Courtesy of Joel Gunn.)

by digital, electrical, or optical techniques. Thus while discrete Fourier transforms can be looked upon (in the above context of analysis of shape in biology) as transformations from the spatial to the frequency domain with the analytical result depicted by clouds of light arranged in polar coordinate form around a central axis, so Good transformations (and their special case, Hadamard/Walsh transforms) and Haar transformations may also be used. Hadamard/Walsh manipulations produce transformations from the spatial to the "sequency" domain, where sequency is a property of rectangular waveforms analogous to the frequency of the sine-cosine waveforms associated with Fourier analysis. In the Haar analysis, the new domain relates to the comparison of points taken in pairs and hence differs from the preceding in that locality information within the spatial domain is revealed. In both the Hadamard/Walsh and the Haar transformations, in contrast to the Fourier analysis, the greatest concentration is in the lower left-hand corner of a square array. Examples of these various transforms as utilized in the processing of pictures taken by the moon landing craft are shown in plates 11 and 12 (Andrews, 1970a, b).

These general methods are very widely applicable throughout science. Such techniques are already well established in a variety of nonbiological studies which include automatic processing of aerial photographs; distinction between ground and cloud patterns in real time; automatic tracking in spark, bubble, and cloud chambers; and automatic signal analysis for explosion monitoring. The

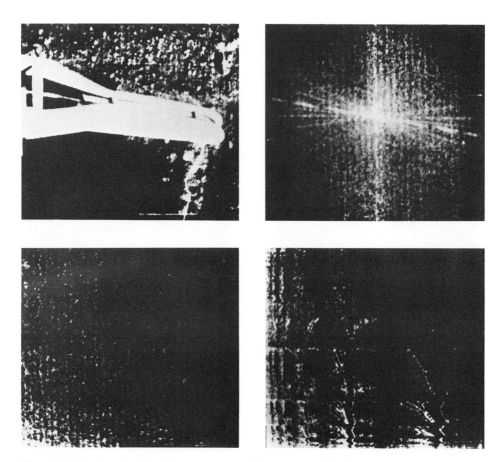

Plate 11. A photograph of the Surveyor spacecraft boom, upper left; the computational Fourier transformation, upper right; Hadamard transformation, lower left; and Haar transformation, lower right. (Courtesy of H. C. Andrews.)

technique of the Walsh transformation has been applied to at least one set of biologically derived data. Meltzer, Searle, and Brown (1967) have utilized it for numerical specification and comparison of leaf forms. The problem that they wish to examine is essentially similar to that presented by comparison of shape in Primates. Thus their use of the Walsh transforms to characterize the shapes of a variety of different leaves from the genus *Althea* (which includes the marshmallow) and *Coleus* (a member of the labiate family of plants) is an attempt to magnify minor variations of shape that differ between specimens as compared with the major shape elements that are similar in all. Their use of the technique is confined to bilaterally symmetrical two-dimensional forms, but the method can be easily generalized to three-dimensional forms and is of special interest because it clearly reveals the small structural features (figure 129). While Meltzer, Searle, and Brown performed this analysis computationally, they recognized the analogy of their histograms to graphs of optical spectra. And within the field of transform

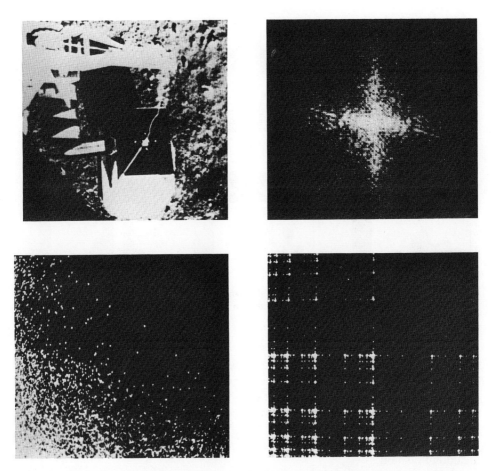

Plate 12. A photograph of the Surveyor spacecraft experimental box, upper left; the logarithms of the magnitudes of: Fourier transformation, upper right; Hadamard transformation, lower left; and Good transformation, lower right. (Courtesy of H. C. Andrews.)

spectroscopy, at least, attempts are being made to produce these types of analyses optically (e.g., Decker, 1970, has constructed and is operating a Hadamard encoding mask).

The use of some of the approaches described in this book is presently a far cry from looking at bones. But I hope that I have been able to demonstrate that there is a great deal of information within bone form and pattern that cannot be obtained by simple observation. For such materials appropriate mathematical transformations, whether performed optically, digitally, or in some other way, may supply a new universe of biologically significant data. However, it is essential that analyses be appropriate; there is little value in applying such methods indiscriminately.

However effective the computers, however involved the mathematics, however intricate the physical equipment, advances in evolutionary studies of primate

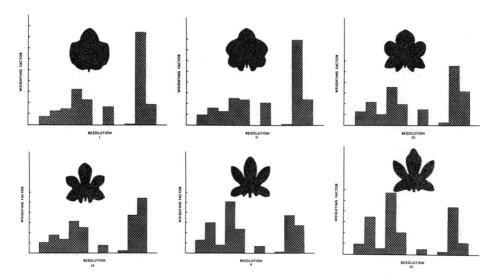

Figure 129. The analysis of leaf form using Walsh functions (After Meltzer, Searle, and Brown, 1967). The higher degrees of resolution are related to increasing amounts of detail on the leaf edges.

morphology cannot be made in the absence of creative anatomical approaches to the original materials. The biological problems must be clearly recognized; where possible, investigators should attempt experimental or pseudo-experimental strategies. Hypotheses should be postulated and tested by at least two independent or semiindependent methods. In all likelihood, little that is worth while remains to be discovered by the technique of drawing upon anatomy in an anecdotal fashion.

An important bonus resulting from the use of methods such as these comes from the sense of community induced within scholars in different areas. This is of especial importance in the study of primates because it is an area which is a confluence for a host of workers. Anatomists, physiologists, and pathologists at the organismal and organ level meet with geneticists, biochemists, and biophysicists at the molecular level, with psychologists, behaviorists, and sociologists at the behavioral grade, and so on. And the volume of work attempted on primates is so large that these various workers continually come into contact. Many of the techniques sketched in these pages are being used in a wide variety of primate studies. Multivariate statistics is employed extensively, especially in human behavioral science, for the investigation of the complex structures of the data universes that are handled. The techniques are also applicable to comparable data from studies of nonhuman primates. Neighborhood limited classification may well be useful as an alternative investigative method in such studies; certainly many other clustering algorithms are proving of value. The techniques of optical data analysis have applications in primate researches that involve the study of complex patterns. Thus Fourier analysis may be used to disentangle the complicated information resulting from the complex interactions of children in playgrounds, patterns of temporal migration of primate troops around waterholes, or complex sequences of facial expressions. Hadamard/Walsh transformations are already being used in making transforms from the time domain to the sequency domain in the analysis of human speech patterns. Equivalent methods could certainly examine the complex wave forms obtained from the sound spectroscopy of primate calls. The whole battery of techniques in pattern recognition, communications, and related disciplines is capable of making both morphological and other data structures yield new orders of information.

References

Abbott, J. C. 1969. *Sets, lattices, and Boolean algebras.* Boston: Allyn and Bacon.

Andrews, H. C. 1970a. Multidimensional rotations in feature selection. In *Symposium on feature extraction and selection in pattern recognition,* ed. S. S. Yau and J. M. Garnett, pp. 9–18. New York: IEEE Computer Group.

———. 1970b. *Computer techniques in image processing.* New York: Academic Press.

Andrews, H. C., and W. K. Pratt. 1969. Transform image coding. In *Symposium on computer processing in communications,* ed. J. Fox, pp. 63–84. Brooklyn: Polytech. Inst.

Ashton, E. H., R. M. Flinn, C. E. Oxnard, and T. F. Spence. 1971. The functional and classificatory significance of combined metrical features of the primate shoulder girdle. *J. Zool.* 163:319–50.

Ashton, E. H., M. J. R. Healy, and S. Lipton. 1957. The descriptive use of discriminant functions in physical anthropology. *Proc. Roy. Soc. B.* 146:552–72.

Ashton, E. H., M. J. R. Healy, C. E. Oxnard, and T. F. Spence. 1965. The combination of locomotor features of the primate shoulder girdle by canonical analysis. *J. Zool.* 147:406–29.

Ashton, E. H., and C. E. Oxnard. 1963. The musculature of the primate shoulder. *Trans. Zool. Soc. Lond.* 29:553–650.

———. 1964. Functional adaptations in the primate shoulder girdle. *Proc. Zool. Soc. Lond.* 142:49–66.

Ashton, E. H., C. E. Oxnard, and T. F. Spence. 1965. Scapular shape and primate classification. *Proc. Zool. Soc. Lond.* 145:125–42.

Attneave, F., and M. D. Arnoult. 1956. The quantitative study of shape and pattern perception. *Psychol. Bull.* 53:452–71.

Avis, V. 1962. Brachiation: the crucial issue for man's ancestry. *Southwest. J. Anthrop.* 18:119–48.

Basmajian, J. V. 1972. Biomechanics of human posture and locomotion: perspectives from electromyography. In *Functional and evolutionary biology of primates: methods of study and recent advances,* ed. R. H. Tuttle, pp. 292–304. Chicago: Aldine-Atherton.

Becker, H. C., P. H. Meyers, and C. M. Nice, Jr. 1969. Laser light diffraction, spatial filtering, and reconstruction of medical radiographic images: preliminary results. *Ann. N.Y. Acad. Sci.* 157:465–86.

Blackith, R. E. 1963. A multivariate analysis of Latin elegiac verse. *Lang. Speech.* 6:196–205.

———. 1965. Morphometrics. In *Theoretical and mathematical biology,* ed. T. H. Waterman and H. J. Morowitz, pp. 225–49. New York: Blaisdell.

Blackith, R. E., and R. M. Blackith. 1969. Variation of shape and of discrete anatomical characters in the morabine grasshoppers. *Aust. J. Zool.* 17:697–718.

Blackith, R. E., and D. K. McE. Kevan. 1967. A study of the genus *Chrotogonus* (Orthoptera). VIII. Patterns of variation in external morphology. *Evol.* 21: 76–84.

Blackith, R. E., and R. A. Reyment, 1971. *Multivariate morphometrics.* London: Academic Press.

Blum, H. 1962. An associative machine for dealing with the visual field and some of its biological implications. In *Biological prototypes and synthetic systems,* Proc. Second. Ann. Bionics Symp. ed. E. E. Bernard and M. R. Kare, pp. 244-60. New York: Plenum.

———. 1967. A transformation for extracting new descriptors of shape. In *Models for the perception of speech and visual form,* Proc. Symp. Data Sciences Lab. and Air Force Cambridge Res. Lab., ed. W. Walthen-Dunn, pp. 362–80. Boston: M. I. T. Press.

Bock, R. D., and E. A. Haggard. 1968. The use of multivariate analysis of variance in behavioral research. In *Handbook of measurement and assessment in the behavioral sciences,* ed. D. K. Whitla, pp. 100–142. Massachusetts: Addison-Wesley.

Boyce, A. J. 1969. Mapping diversity: a comparative study of some numerical methods. In *Numerical taxonomy,* ed. A. J. Cole, pp. 1–30. London: Academic Press.

Bradley, J. V. 1968. *Distribution-free statistical tests.* New Jersey: Prentice-Hall.

Brewster, D. 1816. On the communication of the structure of doubly refracting crystals to glass, muriate of soda, fluorspar and other substances, by mechanical compression and dilatation. *Phil. Trans. Roy. Soc.* 156–78.

Broom, R., and J. T. Robinson. 1950. Further evidence of the structure of the Sterkfontein ape-man *Plesianthropus. Transvaal Mus. Mem.* 4:11–83.

Butler, J. W. 1968. Automatic analysis of bone autoradiographs. In *Pictorial pattern recognition,* ed. G. C. Cheng, pp. 75–85. Washington, D. C.: Thompson.

Campanella, S. J., and G. S. Robinson. 1970. Digital sequency decomposition of voice signals. In *Proceedings of symposium on applications of Walsh functions,* ed. C. A. Bass, pp. 199–202. Washington, D. C.: Naval Research Lab.

———. 1971. A comparison of Walsh and Fourier transformations for application to speech. *IEEE Trans. Electromagnetic Compatibility.* 13:199–202.

Campbell, B. 1937. The shoulder musculature of the platyrrhine monkeys. *J. Mammal.* 18:66–71.

Campbell, B. G. 1966. *Human evolution: an introduction to man's adaptations.* Chicago: Aldine-Atherton.

Carpenter, C. R. 1940. A field study in Siam of the behavior and social relations of the gibbon (*Hylobates lar*). *Comp. Psychol. Mono.* 16:1–212.

Cattell, R. B. 1968. Taxonomic principles for locating and using types (and the derived taxonome computer program). In *Formal representation of human judgement*, ed. B. Kleinmuntz, pp. 99–148. New York: Wiley.

Clark, W. E. Le Gros. 1945. Deformation patterns in the cerebral cortex. In *Essays on growth and form presented to D'Arcy Wentworth Thompson*, ed. W. E. Le Gros Clark and P. B. Medawar, pp. 1–22. Oxford: University Press.

―――――. 1964. *The fossil evidence for human evolution.* 2nd ed. Chicago: Univ. of Chicago Press.

Coker, E. G., and L. N. G. Filon. 1957. *A treatise on photoelasticity.* 2nd ed. rev. H. T. Jessop. Cambridge: University Press.

Cole, A. J., ed. 1969. *Numerical taxonomy.* London: Academic Press.

Currey, J. D. 1968. The adaptation of bones to stress. *J. Theoret. Biol.* 20:91–106.

Dally, J. W., and W. F. Riley. 1965. *Experimental stress analysis.* New York: McGraw-Hill.

Davis, D. D. 1964. The giant panda: a morphological study of evolutionary mechanisms. *Fieldiana Zool. Mem.* 3:1–339.

Davis, J. C. 1970. Optical processing of microporous fabrics. In *Data processing in biology and geology*, Systematics Association Special Volume No. 3., ed. J. L. Cutbill, pp. 69–87. London: Academic Press.

Davis, J. C., and F. W. Preston. 1970. Optical processing: an attempt to avoid the computer. In *Symposium on quantitative geology*, Geological Society of America, special paper.

―――――.1971. Size distributions by optical Fourier analysis. In Proc. Third Int. Congr. Stereol., *Proc. Roy. Micro. Soc.* 6:12–13.

Day, M. H. 1967. Olduvai hominid 10: a multivariate analysis. *Nature, Lond.* 215:323–24.

Day, M. H., and J. R. Napier. 1964. Hominid fossils from Bed I, Olduvai Gorge, Tanganyika: Fossil foot bones. *Nature, Lond.* 201:967–70.

Day, M. H., and B. A. Wood. 1968. Functional affinities of the Olduvai hominid 8 talus. *Man* 3:440–45.

―――――. 1969. Hominoid tali from East Africa. *Nature, Lond.* 222:591–92.

Decker, J. A., Jr. 1970. Experimental Hadamard-transform spectrometry. In *Proceedings of Symposium on applications of Walsh functions*, ed. C. A. Bass, pp. 101–5. Washington, D. C.: Naval Research Lab.

Dobrin, M. B. 1968. Optical processing in the earth sciences. *IEEE Spectrum* 5:59–66.

Dove, R. C., and P. H. Adams. 1964. *Experimental stress analysis and motion measurement.* Columbus: Merrill.

DuBrul, E. L., and D. M. Laskin. 1961. Preadaptive potentialities of the mammalian skull: an experiment in growth and form. *Amer. J. Anat.* 109:117–32.

Duncan, J. P., J. P. Gofton, S. Sikka, and D. Talapatra. 1970. A technique for the topographical survey of biological surfaces. *Med. Biol. Eng.* 8:425–26.

Erikson, G. E. 1954. Comparative anatomy of the New World primates and its bearing on the phylogeny of the apes and man. *Hum. Biol.* 26:210.

Estabrook, G. F. 1966. A mathematical model in graph theory for biological classification. *J. Theoret. Biol.* 12:297–310.

Evans, F. G., and C. W. Goff. 1957. A comparative study of the primate femur by means of the stresscoat and the split-line techniques. *Amer. J. Phys. Anthrop.* 15:59–77.

Fisher, R. A. 1936. The use of multiple measurements in taxonomic problems. *Ann. Eugenics* 7:179–88.

———. 1958. *Statistical methods for research workers.* 13th ed. rev. Edinburgh: Oliver and Boyd.

Flynn, P. D., J. C. Feder, J. T. Gilbert, and A. A. Roll. 1962. Some new techniques for dynamic photoelasticity. *Proc. Soc. Exp. Stress Anal.* 19:159–160.

Frey, H. 1923. Untersuchungen über die Scapula, speziell über ihre äussere Form und deren Abhängigkeit von der Funktion. *Z. Anat. EntwGesch.* 68:277–324.

Frocht, M. M. 1941. *Photoelasticity.* New York: Wiley.

Frost, H. M. 1964. *The laws of bone structure.* Springfield: Thomas.

Gaunt, A. S., and C. Gans. 1969. Mechanics of respiration in the snapping turtle, *Chelydra serpentina* (Linné). *J. Morph.* 128:195–227.

Goodman, J. W. 1968. *Introduction to Fourier optics.* San Francisco: McGraw-Hill.

Gould, S. J. 1969. Character variation in two land snails from the Dutch Leeward Islands: geography, environment and evolution. *Syst. Zool.* 18:185–200.

Gower, J. C. 1967. A comparison of some methods of cluster analysis. *Biometrics* 23:623–36.

Gower, J. C., and G. J. S. Ross. 1969. Minimum spanning trees and single linkage cluster analysis. *Applied Stat. C.* 18:54–64.

Gurdjian, E. S., and H. R. Lissner. 1961. Mechanism of concussion. In *Biomechanical studies of the musculoskeletal system*, ed. F. G. Evans, pp. 192–208. Springfield: Thomas.

Hall, C. E. 1967. *Rotation of canonical variates in multivariate analysis of variance.* Project TALENT, Amer. Inst. Res. and Univ. Pittsburgh.

Hall, A. V. 1969a. Automatic grouping programs: the treatment of certain kinds of properties. *Biol. J. Linn. Soc.* 1:321–25.

———. 1969b. Group forming and discrimination with homogeneity functions. In *Numerical taxonomy*, ed. A. J. Cole, pp. 53–68. London: Academic Press.

Hall-Craggs, E. C. B. 1965. An analysis of the jump of the Lesser Galago (*Galago senegalensis*). *J. Zool.* 147:20–29.

Healy, M. J. R. 1952. Some statistical aspects of anthropometry. *J. Roy. Stat. Soc.* 14:164–84.

———.1968a. Disciplining of medical data. *Brit. Med. Bull.* 24:210–14.

———.1968b. Multivariate normal plotting. *Applied Stat.* 17:157–61.

Hetényi, M. 1950. *Handbook of experimental stress analysis.* New York: Wiley.

Hiernaux, J. 1963. Heredity and environment: their influence on human morphology. A comparison of two independent lines of study. *Amer. J. Phys. Anthrop.* 21:575–89.

Hildebrand, M. 1967. Symmetrical gaits of primates. *Amer. J. Phys. Anthrop.* 26:119–30.

Holeman, J. M. 1968. Holographic character reader. In *Pattern recognition*, ed. L. N. Kanal, pp. 63–78. Washington, D. C.: Thompson.

Holister, G. S. 1961. Recent developments in photoelastic coating techniques. *J. Roy. Aeron. Soc.* 65:661–69.

———. 1967. *Experimental stress analysis.* Cambridge: University Press.

Howells, W. W. 1951. Factors of human physique. *Amer. J. Phys. Anthrop.* 9:159–91.

———. 1957. The cranial vault: factors of size and shape. *Amer. J. Phys. Anthrop.* 15:19–48.

———. 1966. Craniometry and multivariate analysis. The Jomon population of Japan. A study by discriminant analysis of Japanese and Ainu crania. *Papers Peabody Mus. Arch. Ethnol., Harvard Univ.* 57:1–43.

———. 1968. Measurement and analysis in anthropology. In *Handbook of measurement and assessment in the behavioral sciences*, ed. D. K. Whitla, pp. 393–418. Massachusetts: Addison-Wesley.

———. 1969. The use of multivariate techniques in the study of skeletal populations. *Amer. J. Phys. Anthrop.* 31:311–14.

———. 1972. Analysis of patterns of variation in crania of recent man. In *Functional and evolutionary biology of primates: methods of study and recent advances*, ed. R. H. Tuttle, pp. 123–51. Chicago: Aldine-Atherton.

Huxley, J. S. 1932. *Problems of relative growth.* London: Methuen.

Inman, V. T., J. B. De C. M. Saunders, and LeR. C. Abbott. 1944. Observations on the function of the shoulder joint. *J. Bone and Jt. Surg.* 26:1–30.

Jardine, N., and R. Sibson. 1971. *Mathematical taxonomy.* London: Wiley.

Jenkins, F., Jr. 1970. Limb movements in a monotreme (*Tachyglossus aculeatus*): a cineradiographic analysis. *Science* 168:1473–75.

Johnston, F. E., J. M. Tanner, R. W. Whitehouse, and P. H. A. Sneath. 1967. A mathematical analysis of shape changes in growing children. *Amer. J. Phys. Anthrop.* 27:246

Justus, R., and J. H. Luft. 1970. A mechanochemical hypothesis for bone remodeling induced by mechanical stress. *Calc. Tiss. Res.* 5:222–35.

Kummer, B. 1959. *Bauprinzipien des Säugerskeletes.* Stuttgart: Thieme.

Lance, G. N., and W. T. Williams. 1966. A generalized sorting strategy for computer classifications. *Nature, Lond.* 212:218.

Leakey, L. S. B., P. V. Tobias, and J. R. Napier. 1964. A new species of the genus *Homo* from Olduvai Gorge. *Nature, Lond.* 202:7–9.

Leven, M. M. 1955. Quantitative three-dimensional photoelasticity. *Proc. Soc. Exp. Stress Anal.* 12:157–71.

Lewis, O. J. 1971. Brachiation and the early evolution of the hominoidea. *Nature, Lond.* 230:577–79.

———. 1972. The evolution of the hominoid wrist. In *Functional and evolutionary biology of primates: methods of study and recent advances,* ed. R. H. Tuttle, pp. 207–22. Chicago: Aldine-Atherton.

Liem, K. F. 1970. Comparative functional anatomy of the Nandidae (Pisces: Teleostei). *Fieldiana Zool. Mem.* 56:1–166.

Lisowski, F. P. 1967. Angular growth changes and comparisons in the primate talus. *Folia Primatol.* 7:81–97.

Meltzer, B., N. H. Searle and R. Brown. 1967. Numerical specification of biological form. *Nature, Lond.* 216:32–36.

Merriam, D. F., and J. W. Harbaugh. 1964. Trend surface analysis of regional and residual components of geologic structures in Kansas. *Kansas St. Geol. Surv. Spec. Distr. Publ.* No. 11, pp. 1–27.

Miller, R. A. 1932. Evolution of the pectoral girdle and forelimb in primates. *Amer. J. Phys. Anthrop.* 17:1–56.

Moss, W. W. 1968. Experiments with various techniques of numerical taxonomy. *Syst. Zool.* 17:31–47.

Mukherjee, R., C. R. Rao, and J. C. Trevor. 1955. *The ancient inhabitants of the Jebel Moya.* Cambridge: University Press.

Murray, P. D. F. 1936. *Bones. A study of the development and structure of the vertebrate skeleton*. Cambridge: University Press.

Napier, J. R. 1959. Problem of brachiation among primates with special reference to *Proconsul. Ber 6. Tagung dtsch. Ges Anthrop.* 28.

———. 1961. Prehensility and opposability in the hands of primates. *Symp. Zool. Soc. Lond.* 5:115–32.

———. 1962. Fossil hand bones from Olduvai Gorge. *Nature, Lond.* 196:400–411.

Napier, J. R. and P. R. Davis. 1959. The forelimb skeleton and associated remains of Proconsul africanus. *Fossil mammals of Africa*. No. 16: pp. 1–78.

Nathan, R., and R. H. Selzer. 1968. Digital video data handling: Mars, the moon and men. In *Image processing in biological science,* ed. D. M. Ramsey, pp. 177–210. Los Angeles: Univ. of California.

Neely, P. M. 1972. Towards a theory of classification. I. Neighborhood limited classification. II. A consistent locally stable classificatory algorithm. Unpublished manuscripts.

Oxnard, C. E. 1961. The forelimb in primates. *Abstr. Soc. Study Hum. Biol.* 7:1.

———. 1963. Locomotor adaptations in the primate forelimb. *Symp. Zool. Soc. Lond.* No. 10; pp. 165–82.

———. 1966. Some functional osteometric features of the primate innominate bone. *Proc. Int. Primatol. Soc. Frankfurt.* 59–60.

———. 1967. The functional morphology of the primate shoulder as revealed by comparative anatomical, osteometric and discriminant function techniques. *Amer. J. Phys. Anthrop.* 26:219–40.

———. 1968a. The architecture of the shoulder in some mammals. *J. Morph.* 126:249–90.

———. 1968b. A note on the fragmentary Sterkfontein scapula. *Amer. J. Phys. Anthrop.* 28:213–18.

———. 1968c. A note on the Olduvai clavicular fragment. *Amer. J. Phys. Anthrop.* 29:429–32.

———. 1969a. Aspects of the mechanical efficiency of the hands of higher primates as demonstrated by two-dimensional photoelasticity. *Anat. Rec.* 163: 239.

———. 1969b. Mathematics, shape and function: a study in primate anatomy. *Amer. Sci.* 57:75–96.

———. 1969c. The combined use of multivariate and clustering analyses in functional morphology. *J. Biomech.* 2:73–88.

———. 1969d. Evolution of the human shoulder: some possible pathways. *Amer. J. Phys. Anthrop.* 30:319–31.

———. 1970. Computer methods and functional morphology in primates. *VIII Congr. Anthr. Ethnol. Sci.* S 3:324–27.

———. 1971. Tensile forces in skeletal structures. *J. Morph.* 134:425–36.

———. 1972a. Functional morphology of primates: some mathematical and physical methods. In *Functional and evolutionary biology of primates: methods of study and recent advances,* ed. R. H. Tuttle, pp. 305–36. Chicago: Aldine-Atherton.

———. 1972b. Some African fossil foot bones: A note on the interpolation of fossils into a matrix of extant species. *Amer. J. Phys. Anthrop.* 37:3–12.

———. 1972c. Some problems in the comparative assessment of skeletal form. In *Symposium on human evolution,* ed. M. H. Day, pp. 1–23. London: Taylor and Francis.

———. 1972d. The use of optical data analysis in functional morphology: investigation of vertebral trabecular patterns. In *Functional and evolutionary biology of primates: methods of study and recent advances,* ed. R. H. Tuttle, pp. 337–47. Chicago: Aldine.

Oxnard, C. E., and P. M. Neely. 1969. The descriptive use of neighborhood limited classification in functional morphology. *J. Morph.* 129:1–22.

Pauwels, F. 1948. Die Bedeutung der Bauprinzipien des Stütz-und Bewegungsapparates für die Beanspruchung der Röhrenknochen. Erster Beitrag zur funktionellen Anatomie und kausalen Morphologie des Stützapparates. *Z.Anat. EntwGesch.* 114:129–66.

———. 1965. *Gesammelte Abhandlungen zur funktionellen Anatomie des Bewegungsapparates.* Berlin: Springer-Verlag.

Philbrick, O. 1968. Shape description with the medial axis transformation. In *Pictorial pattern recognition,* ed. G. C. Cheng, R. S. Ledley, D. K. Pollack, and A. Rosenfeld, pp. 395–407. Washington, D. C.: Thompson.

Pincus, H. J. 1969. Sensitivity of optical data processing to changes in rock fabric. *Int. J. Rock Mech. Min. Sci.* 6:259–76.

Preston, D. A. 1970. Fortran IV program for sample normality tests. *Kansas St. Geol. Surv. Computer contribution* 41:1–28.

Preston, F. W., D. W. Green, and J. C. Davis. 1969. Numerical characterization of reservoir rock pore structure. *Second Ann. Rep. to the Amer. Petrol. Inst. Research Project no. 103,* pp. 1–84.

Preuschoft, H. 1970. Functional anatomy of the lower extremity. In *The chimpanzee,* vol. 3, ed. G. H. Bourne, pp. 221–94. Basel: Karger.

———. 1971. Body posture and mode of locomotion in early pleistocene hominids. *Folia Primatol.* 14:209–40.

Quenouille, M. H. 1952. *Associated measurements.* London: Butterworth.

Rosenfeld, A. 1969. *Picture processing by computer.* New York: Academic Press.

Ross, G. J. S. 1969. Single linkage cluster analysis. *Applied Stat.* 18:106–10.

Rothman, R. H. 1967. Electrical and mechanical principles in bone dynamics. In *Engineering in the practice of medicine*, ed. B. L. Segal and D. G. Kilpatrick, pp. 142–47. Baltimore: Williams and Wilkins.

Rubin, J., and H. P. Friedman. 1967. *A cluster analysis and taxonomy system for grouping and classifying data.* New York: I. B. M. Corporation.

Salmons, S. 1969. The eighth International Conference on Medical and Biological Engineering. *Bio-Med. Eng.* 4:467–74.

Seal, H. 1964. *Multivariate statistical analysis for biologists.* London: Methuen.

Searle, S. R. 1966. *Matrix algebra for the biological sciences.* New York: Wiley.

Siegel, S. 1956. *Nonparametric statistics for the behavioral sciences.* New York: McGraw-Hill.

Selzer, R. H. 1968. Improving biomedical image quality with computers. NASA Tech. Rep. 32-1336. Pasadena: Jet Propulsion Lab., Calif. Inst. Tech.

Shelman, C. B., and D. Hodges. 1970. A general purpose program for the extraction of physical features from a black and white picture. In *Symposium on feature extraction and selection in pattern recognition*, ed. S. S. Yau and J. M. Garnett, pp. 135–44. New York: IEEE Computer Group.

Shulman, A. R. 1970. *Optical data processing.* New York: Wiley.

Simpson, G. G. 1945. The principles of classification and a classification of mammals. *Bull. Amer. Mus. Nat. Hist.* 85:1–350.

Singer, F. L., H. Milch, and R. A. Milch. 1964. Distribution of surface strain in paired human femora. *Nature, Lond.* 202:206–8.

Smith, J. W. 1962. The relationships of epiphysial plates to stress in the bones of the lower limb. *J. Anat., Lond.* 96:58–78.

Sneath, P. H. A. 1967. Trend-surface analysis of transformation grids. *J. Zool., Lond.* 151:65–122.

Snedecor, G. W., and W. G. Cochran. 1967. *Statistical methods.* 5th ed. Iowa: Iowa State University.

Sokal, R. R., and P. H. A. Sneath. 1963. *Principles of numerical taxonomy.* San Francisco: Freeman.

Sokal, R. R., and F. J. Rohlf. 1969. *Biometry: the principles and practice of statistics in biological research.* San Francisco: Freeman.

Stern, J. T., Jr. 1970. The meaning of "adaptation" and its relation to the phenomenon of natural selection. In *Evolutionary biology*, vol. 4, ed. Th. Dobzhansky, M. K. Hecht, and W. C. Steere, pp. 39–66. New York: Appleton-Century-Crofts.

———. 1971a. Functional myology of the hip and thigh of cebid monkeys and its implications for the evolution of erect posture. *Bibliotheca Primatol.* 14:1–318.

---. 1971b. Investigations concerning the theory of "spurt" and "shunt" muscles. *J. Biomech.* 4:437–53.

Stern, J. T., Jr., and C. E. Oxnard. 1973. *Primate locomotion: some links with evolution and morphology.* Handbook of Primatology. Basel: Karger.

Tanner, J. M., F. E. Johnston, R. W. Whitehouse, and P. H. A. Sneath. 1969. Changes in the shape of individual children studied longitudinally from 5 to 20 years of age. *Hum. Biol.* 41:282–83.

Tappen, N. C. 1960. Problems of distribution and adaptation of the African monkeys. *Curr. Anthrop.* 1:91–120.

Theile, F. W. 1884. Gewichtsbestimmungen zur Entwickeburg der Skelettes beim Menschen. *Nova Acta Leop. Carol.* 46:134–471.

Theocaris, P. S. 1969. *Moiré fringes in strain analysis.* Oxford: Pergamon.

Thom, R. 1970. Topological models in biology. In *Towards a theoretical biology*, 3. Drafts, ed. C. H. Waddington, pp. 89–116. Chicago: Aldine.

Thomas, F. D., and E. E. Sellers, eds. 1969. *Biomedical sciences instrumentation*, vol. 6. Pittsburgh: Instrument Society of America.

Thompson, D'Arcy W. 1917, 1942. *On growth and form.* Cambridge: University Press.

Timoshenko, S. 1955. *Strength of materials. Part I: Elementary theory and problems.* Princeton: Van Nostrand.

Tobias, P. V. 1967. The cranium and maxillary dentition of *Australopithecus* (*Zinjanthropus*) *boisei.* In *Olduvai Gorge*, vol. 2, ed. L. S. B. Leakey. Cambridge: University Press.

Tolles, W. E., ed. 1969. Data extraction and processing of optical images in the medical and biological sciences. *Ann. N.Y. Acad. Sci.* 157:1–530.

Tukey, J. W. 1962. The future of data analysis. *Annals Math. Stat.* 33:1–67.

---. 1970. Some further inputs. In *Geostatistics,* ed. D. F. Merriam, pp. 163–74. New York: Plenum.

Tuttle, R. H. 1967. Knuckle-walking and the evolution of hominoid hands. *Amer. J. Phys. Anthrop.* 26:171–206.

---. 1969. Quantitative and functional studies on the hands of the Anthropoidea. I. The Hominoidea. *J. Morph.* 128:309–64.

---. 1972. Relative mass of cheiridia muscles in catarrhine primates. In *Functional and evolutionary biology of primates: methods of study and recent advances,* ed. R. H. Tuttle. pp. 262–91. Chicago: Aldine-Atherton.

Uhlman, K. 1969. Huft-und Oberschenkelmuskulatur Systematische und Vergleichende Anatomie. *Primatologia* 4:1–442.

Underwood, E. E. 1970. *Quantitative stereology.* Massachusetts: Addison-Wesley.

Wagstaffe, W. W. 1874. On the mechanical structure of the cancellous tissue of bone. *St. Thomas Hosp. Rep.* n.s. 5:192–214.

Washburn, S. L. 1947. Relation of the temporalis muscle to the form of the skull. *Anat. Rec.* 99:239–48.

———. 1964. Behavior and human evolution. In *Classification and human evolution,* ed. S. L. Washburn, pp. 190–203. Chicago: Aldine.

———. 1967. Behavior and the origin of man. Huxley Memorial Lecture. *Proc. Roy. Anthro. Inst.* 21–27.

———. 1968. The study of human evolution. Condon Lectures. *Oregon State System of Higher Education,* Eugene, Oregon. 1–48.

Wilk, M. B., and R. Gnanadesikan. 1968. Probability plotting methods for the analysis of data. *Biometrika* 55:1–17.

Winsor, F. 1958. *The space child's mother goose.* New York: Simon and Schuster.

Wishart, D. 1969. Mode analysis: a generalization of nearest neighbor which reduces chaining effects. In *Numerical taxonomy,* ed. A. J. Cole, pp. 282–308. London: Academic Press.

Woodger, J. H. 1945. On biological transformations. In *Essays on growth and form presented to D'Arcy Wentworth Thompson,* ed. W. E. Le Gros Clark and P. B. Medawar, pp. 95-120. Oxford: University Press.

Yau, S. S. and J. M. Garnett, eds. 1970. *Conference record of the symposium on feature extraction and selection in pattern recognition.* New York: IEEE Computer Group.

Ziegler, A. C. 1964. Brachiating adaptations of chimpanzee upper limb musculature. *Amer. J. Phys. Anthrop.* 22:15–31.

Zuckerman, S. 1966. Myths and methods in anatomy. *J. Roy. Coll. Surg. Edin.* 2:87–114.

———. 1970. *Beyond the ivory tower.* London: Weidenfeld and Nicolson.

Zuckerman, S., E. H. Ashton, C. E. Oxnard, and T. F. Spence. 1967. The functional significance of certain features of the innominate bone in living and fossil primates. *J. Anat. Lond.* 101:608–9.

Author Index

Abbott, J. C., 94
Abbott, LeR. C., 26, 27
Adams, P. H., 123
Anderson, E., 6–10
Andrews, H. C., 175, 197, 199
Arnoult, M. D., 11, 12
Ashton, E. H., 21, 26, 30, 35, 37–38, 50–52, 54, 67, 88, 107, 112, 145, 152
Attneave, F., 11, 12
Avis, V., 20

Basmajian, J. V., 27, 122
Becker, H. C., 174, 192
Blackith, R. E., 37–38, 48–49
Blum, H., 12, 13, 169
Bock, R. D., 37
Boyce, A. J., 112–13
Bradley, J. V., 87
Brewster, D., 16
Broom, R., 159
Brown, R., 198, 200
Butler, J. W., 175

Campanella, S. J., 175
Campbell, B., 21
Campbell, B. G., 21, 158
Carpenter, C. R., 20
Cattell, R. B., 89
Clark, W. E. LeG., 16, 152, 171
Cochran, W. G., 87
Coker, E. G., 123
Cole, A., 2
Currey, J. D., 136

Dally, J. W., 123
Davis, D. D., 88
Davis, J. C., 175, 182–84, 193
Davis, P. R., 21
Day, M. H., 159–62, 165
Decker, J. A., Jr., 199
Dobrin, M. B., 191, 193
Dove, R. C., 123
DuBrul, E. L., 111
Duncan, J. P., 123

Erickson, G. E., 21
Estabrook, G. F., 94
Evans, F. G., 123

Feder, J. T., 124
Filon, L. N. G., 123
Fisher, R. A., 1, 6, 8, 87, 101
Flinn, R. M., 21, 38, 52, 54, 67, 107, 112, 152

Flynn, P. D., 124
Frey, H., 26, 67
Friedman, H. P., 7, 10, 120
Frocht, M. M., 123
Frost, H. M., 136, 144

Gans, C., 122
Garnett, J. M., 175
Gaunt, A. S., 122
Gibbs, C. B., 70
Gilbert, J. T., 124
Gnanadesikan, R., 87
Goff, C. W., 123
Gofton, J. P., 123
Goodman, J. W., 179, 195
Gould, S. J., 42
Gower, J. C., 112, 113
Green, D. W., 175, 182
Gunn, J. D., 196–97
Gurdjian, E. S., 123

Haggard, E. A., 37
Hall, A. V., 113
Hall, C. E., 56
Hall-Craggs, E. C. B., 20
Harbaugh, J. W., 11
Healy, M. J. R., 21, 37, 50, 69–70, 82–83, 85, 88, 145
Hetényi, M., 123
Hiernaux, J., 49
Hildebrand, M., 20
Hodges, D., 11–12, 175
Holeman, J. M., 188
Holister, G. S., 123–24
Howells, W. W., 37, 47, 49
Huxley, J. S., 1

Inman, V. T., 26, 67

Jardine, N., 111
Jenkins, F., Jr., 122
Johnston, F. E., 11
Justus, R., 144

Kevan, D. K. McE., 37, 48
Kummer, B., 124

Lance, G. N., 112
Laskin, D. M., 111
Leakey, L. S. B., 160
Leven, M. M., 124
Lewis, O. J., 155, 168
Liem, K. F., 111, 122
Lipton, S., 37, 88
Lissner, H. R., 123
Lisowski, F. P., 165
Luft, J. H., 144

McPherron, A., 196
Meltzer, B. N., 198, 200
Merriam, D. F., 11
Meyers, P. H., 174, 192
Milch, H., 123
Milch, R. A., 123
Miller, R. A., 26–27
Moss, W. W., 112–13
Mukherjee, R., 2, 37
Murray, P. D. F., 21, 171

Napier, J. R., 21, 134, 160
Nathan, R., 174
Neely, P. M., 8, 10, 21, 94, 101, 104–5
Nice, C. M., Jr., 174, 192

Oxnard, C. E., 21–22, 26, 30, 33, 35, 37–38, 50–54, 58, 66–67, 71, 94, 104–5, 107, 111–13, 128, 138, 145, 148, 152, 155–57, 160, 162–64, 169, 183

Pauwels, F., 16–17, 124
Philbrick, O., 169
Pincus, H. J., 181
Pratt, W. K., 175
Preston, F. W., 87, 175, 182
Preuschoft, H., 124, 168

Quenouille, M. H., 81, 87

Rao, C. R., 2, 37
Reyment, R. A., 38, 48
Riley, W. F., 123
Robinson, G. S., 175
Robinson, J. T., 159
Rohlf, F. J., 87
Roll, A. A., 124
Rosenfeld, A., 14, 175
Ross, G. J. S., 112
Rothman, R. H., 21
Rubin, J., 7, 10, 120

Salmons, S., 122
Saunders, DeC. M., 26, 67
Seal, H., 38
Searle, N. H., 198, 200
Searle, S. R., 38
Sellers, E. E., 175
Selzer, R. H., 174, 193
Shelman, C. B., 11–12, 175

Shulman, A. R., 174–75, 182, 191, 194–95
Sibson, R., 111
Siegel, S., 87
Sikka, S., 123
Simpson, G. G., 19
Singer, F. L., 123
Smith, J. W., 124
Sneath, P. H. A., 1–2, 11, 112
Snedecor, G. W., 87
Sokal, R. R., 2, 87, 112
Spence, T. F., 21, 35, 37–38, 50–52, 54, 67, 107, 112, 145, 152
Stern, J. T., Jr., 22, 25, 34–35, 122

Talapatra, D., 123
Tanner, J. M., 11
Tappen, N. C., 110
Theile, F. W., 76
Theocaris, P. S., 123
Thom, R., 1
Thomas, F. D., 175
Thompson, D'A. W., 1, 11
Timoshenko, S., 127
Tobias, P. V., 6, 160
Tolles, W. E., 175
Tonkinson, T., 142
Trevor, J. C., 2, 37
Tukey, J. W., 69, 82
Tuttle, R. H., 20–21, 31–32, 76, 127

Uhlman, K., 34
Underwood, E. E., 172

Wagstaffe, W. W., 171
Washburn, S. L., 21, 36, 155
Whitehouse, R. W., 11
Wilk, M. B., 87
Williams, W. T., 112
Wishart, D., 112
Wood, B. A., 161–62, 165
Woodger, J. H., 1

Yau, S. S., 175

Ziegler, A. C., 67
Zuckerman, S., 35, 67, 152

Species Index

Ailuropoda (giant panda), 88
Alouatta (howler monkey), 29, 35, 56, 67, 75, 103–5, 110–11, 115
Althea (marshmallow), 158
Angwantibo. See *Arctocebus*
Anomalurus. See Squirrel, "flying"
Anteater, 57, 60
Anthropoidea, 35, 54–55, 103, 106–9, 117, 157
Aotus, Douroucouli (night monkey), 103, 115
Apes, 31–32, 34–35, 52, 67–68, 102, 104, 127, 141, 155–57, 159–65, 168, 174
Arboreal mammals, 57–62, 64–65
Arctocebus (angwantibo), 30, 35, 52–54, 56, 58, 105
Ateles (spider monkey), 27, 29, 102–4, 110, 115, 155–56
Atelines, 67
Australopithecines, 152, 161–64
Australopithecus, 68, 155, 157, 160, 165, 168
Aye-aye. See *Daubentonia*

Babirussa, 73–74
Baboon. See *Papio*
Bats. See *Chiroptera*
Brachyteles (woolly spider-monkey), 53–54, 56, 72, 102–4, 115
Bushbaby. See *Galago*

Cacajao (uakari monkey), 53–57, 59, 103, 110, 115, 155–58
Callicebus (titi monkey), 103, 115
Callimico (Goeldi's marmoset), 103, 115
Calliosciurus, 72
Callithrix (marmoset), 67, 103, 115
Capuchin. See *Cebus*
Carnivores, 64
Ceboidea, 146, 149–51
Cebus (capuchin), 59, 103, 115
Cercocebus (mangabey), 74–75, 102–4, 110, 115
Cercopithecines, 35
Cercopithecoidea, 146, 149–51

Cercopithecus (guenons), 27, 29–30, 35, 57, 72, 74–75, 102–3, 107, 110, 115
Chiropotes, 110
Chiroptera (bats), 61–62, 66, 72, 138, 140
Chrotogonus (grasshopper), 48
Coleus, 198
Colobines, 35
Colobus (guereza), 29, 56, 103, 115

Daubentonia (aye-aye), 52–53, 72, 110, 115, 117, 145–51
Dermopterans, 60–61, 64–66
Desmodus, 72
Douroucouli. See *Aotus*

Edentates, 60, 64–65
Erinaceus, 72
Erythrocebus (patas monkey), 30, 35, 52–54, 56–57, 76, 102–3, 117, 155–58
Euoticus (needle-clawed bushbaby), 52–54, 58–59

Fat-tailed lemur. See *Hapalemur*
"Flying" lemurs. See Indriids
"Flying" phalanger, 61

Galagines, 67–68
Galago, 35, 52–54, 56, 58
Gibbon. See *Hylobates*
Goeldi's marmoset. See *Callimico*
Gorilla, 22–24, 29, 31–32, 54–57, 63, 67–68, 72, 76, 102–4, 111, 115, 118–21, 131, 133, 142, 155–59, 162
Guenon. See *Cercopithecus*
Guereza. See *Colobus*

Hapalemur (fat-tailed lemur), 52–54, 58
Hominoidea, 31, 35, 116, 118–19, 146, 149–51
"*Homo habilis*," Olduvai, 155, 157, 159–60, 168
Howler monkey. See *Alouatta*
Hylobates (gibbon), 22–24, 27, 29, 31–32, 52–54, 59, 68, 102–4, 110–11, 114–15, 117–21, 133, 137, 156–57
Hylobatines, 31–32

Indriids, 54, 67
Iris, 6, 7–10, 99

Jerboa, 60

Kangaroo, 60
Kromdraai talus, 160–64

Lagothrix (woolly monkey), 27, 29–30, 53–54, 56, 102–5, 115, 156
Langur. See *Presbytis*
Lemur, 35, 52–54, 58–59, 107, 109
Lemuroidea, 146, 149–51
Leontocebus (tamarin), 72, 103, 115
Lepilemur (weasel lemur), 52–54, 56, 58
Loris (slender loris), 52–54, 58
Lorisiformes, 146, 149–51
Lorisines, 54, 67–68

Macaca, 28, 54–55, 57, 74–75, 102–3, 110, 115, 133, 137
Man, 23–25, 28, 31, 34–35, 50, 52–59, 61–62, 64, 66–67, 72, 103, 114, 116–21, 127, 133, 136, 152–68, 171–74
Mandrillus, 102–3, 115, 117
Mangabey. See *Cercocebus*
Marmoset. See *Callithrix*
Marsupials, 60, 64–66
Mirid bug. See *Plagognathus*

Nasalis (proboscis monkey), 27, 29–30, 54–55, 102–3, 115, 156–57
Neandertals, 160–64
Needle-clawed bushbaby. See *Euoticus*
New World monkeys, 34–35, 52, 67, 102–3, 105, 110, 114–18, 156
Night monkey. See *Aotus*
Non-primate mammals, 56–62, 64–65
Nycticebus (slow loris), 52–54, 58

Olduvai: clavicle, 155; hand, 134, 164–67; hominid 10, 159–60, 165; proximal phalanx, 165–67; talus, 161–65

Old World monkeys, 52, 67–68, 102–3, 105, 110, 114–17
Orangutan. See *Pongo*

Pan (chimpanzee), 29, 31–32, 34, 54–57, 67–68, 72, 103–4, 118–21, 128–32, 135, 152–60, 162, 165–67, 171–74
Papio (baboon), 27, 29, 54, 56–57, 59, 63, 65, 102–4, 111, 115, 117, 137, 155–58
Patas monkey. See *Erythrocebus*
Perodicticus (potto), 22, 30, 35, 52–54, 58–59, 105
Petaurus, 72
Pithecia (saki monkey), 34, 54–55, 57, 65, 103, 110, 115, 155–58
Plagognathus (mirid bug), 48
Pongo (orangutan), 23, 28–29, 31–33, 35, 54–57, 59, 68, 72, 76, 88, 103–4, 114–16, 118–21, 128–29, 131–32, 134–35, 155–59, 165–67
Potto. See *Perodicticus*
Prehensile-tailed monkeys, 34
Presbytis (langurs), 53, 59, 74–75, 103, 115, 158;
high-canopy, 54–55, 57, 75, 110, 155–57; low-canopy, 54, 57, 75, 104, 110, 155–57
Proboscis monkey. See *Nasalis*
Proconsul talus, 161–65
Procyonids, 88
Propithecus (sifaka), 35, 52–54, 58
Prosimii, 22, 35, 51–56, 58, 67, 105–10, 115–17
Pygathrix, 103, 110, 115
Pygmy chimpanzee, 72
Pygmy, human, 152, 172–74

Rhinopithecus (snub-nosed langur), 57, 66, 103–4, 115, 155
Rodents, 60, 64–66

Saimiri (squirrel monkey), 27, 29–30, 35, 52–53, 102–3, 107, 115
Saki monkey. See *Pithecia*
Siamang. See *Symphalangus*
Sifaka. See *Propithecus*
Simias, 110
Slender loris. See *Loris*

Sloth, 57, 60
Slow loris. See *Nycticebus*
Snub-nosed langur. See *Rhinopithecus*
Sphaeronycteris, 72
Spider monkey. See *Ateles*
Squirrels; "flying," 57, 60–61; ground, 57, 60; tree, 57, 60–62
Squirrel monkey. See *Saimiri*
Sterkfontein: femur, 168; innominate, 151–54, 165, 170–74; scapula, 155; terminal toe phalanx, 168
Symphalangus (siamang), 29, 104, 110

Tadarida, 72
Tamarin. See *Leontocebus*
Titi monkey. See *Callicebus*
Tree shrew. See *Tupaia*
Tupaia (tree shrew), 107–9

Uakari. See *Cacajao*

Weasel lemur. See *Lepilemur*
Woolly monkey. See *Lagothrix*
Woolly spider-monkey. See *Brachyteles*

Subject Index

Acetabulum, 35
Acromion, 30
Aggregated groups, 89–90, 93, 108–9
Allometry, 1, 6, 16, 152
Amplitude spectrum, 178, 188
Analysis of pattern, 171–201
Analyzer, 125
Anisotropic data, 92–94, 107
Anisotropic properties of bone, 124
Archaeological artefacts, 196–97
Auricular facet, 35

Barycentric decomposition, 96
Biceps muscle, 159
Biological information in mathematical parameters, 48–51, 55, 62–67, 102–3, 133, 135
Bivariate analysis, 6–7, 40, 43, 83–86
Boundary stress, 125, 143
Brachialis muscle, 17
Brittle-lacquer coatings, 122–23
Buckle transducer, 122

Canonical analysis, 37–38, 44–46, 48, 88, 105, 107, 109; in combination with group-finding procedures, 111–21; contribution of original dimensions to, 62–66; interpolative studies, 145–51; of pelvic dimensions, 151–55; of shoulder dimensions, 50–68, 155–59; of talus dimensions, 160–64; of terminal toe phalanx dimensions, 159–60
Cartesian grids. *See* Coordinate measurements
Cluster-finding. *See* Group-finding procedures
Coordinate measurements, 1, 11, 63
Coracobrachialis muscle, 17
Coracoid process, 159
Correlation, 6, 16, 31, 37–39, 79
Crossed polariscope: circular, 124–26; plane, 125–26
Cube root transformation, 85
Cumulative normal plotting, 82–83

Deltoid muscle, 27, 30, 138
Dendrogram, 113, 115, 117
Digital flexor muscles, 31, 33–34, 128, 134
Digital ray, 127–28, 130–36, 165–68
Digitized patterns, 174, 176
Discriminant function analysis, 6, 8, 10, 47, 121
Distribution: bimodal, 74; kurtotic, 81; non-normal, 14, 77–79, 81; normal, 8, 78, 80, 82; skewed, 74, 79–81

Eigenvector space. *See* Group-finding procedures
Electromyography, 16, 122, 138
Experimental stress analysis, 16, 63, 164

Factor analysis, 37, 38, 42, 47, 88
Fast Fourier transforms, 14, 177
Femur, 141, 168
Flux of light, 178–80
Flying spot scanners, 11, 171, 174–75
Fourier analysis, 14, 175–201; one-dimensional, 176–77; power spectrum, 15, 180, 182–90, 192, 195–97; two-dimensional, 177–78
Frozen stress technique, 124

Glenoid cavity, 30–31, 159
Generalized distance analysis, 37–38, 44, 47–58, 85–86, 88; of *Daubentonia* data, 147–51; of shoulder data, 112–18; of talus data, 161–63, 165; of terminal toe phalanx data, 160
Good transformation, 197, 199
Group-finding procedures, 7–10, 48, 88–121, 163–64, 200; combined with multivariate methods, 111–18, 147–48; in eigenvector space, 7–8, 10, 120, 121; intrageneric, 110; suprageneric, 109

Haar transformation, 197–98
Hadamard transformation. *See* Walsh transformation
Hand, 21, 23–24, 27, 31–34, 168; hook-like function of, 23, 33, 129, 131, 134, 136, 165, 167
"Hanging rootogram," 82
Histogram, 78–85
Holography, two-dimensional, 188
Humerus, 17

IDECS, 180, 181, 190
Ilium, 35, 152–53, 171
Image enhancement, 181
Interpolation, 16–17, 118, 145–68
Ischium, 35, 142, 152, 154, 171
Isochromatics, 17, 124–27, 130, 132, 135–37, 166–67
Isoclinics, 126–27
Isotropic data, 91–94

Knee, 124, 141–42, 168
Knuckle-walking, 23, 31–34, 128–32, 136, 155, 165–68

Laser, 175–76, 188; contouring, 196
Latin elegaic verse, 49
Latissimus dorsi muscle, 29–30
Logarithmic transformation, 83, 85, 93, 199

Maximum shearing stress, 125
Mechano-electric properties, 16, 21
Medial axis transformation, 12–13, 158, 169–74
Metacarpal, 33, 128, 130, 132
Metacarpophalangeal joints, 32–33
Minimum spanning tree, 112–14, 116–18, 147–48
Misclassification, 8, 10
Moiré fringe analysis: contouring, 169, 196; strain analysis, 122–23
Monochromatic light, 125
Morphogenetic channels, 25, 109–11
Morphologic peaks, 109–11

Subject Index

Multivariate normal plotting, 85–86
Muscle weights, 29, 32, 34, 76

Neighborhood limited classification, 8, 10, 94, 97–103, 119–20, 163–64, 200; compared with canonical analysis, 118–21; compared with grouping in eigenvector space, 163–64; of shoulder data, 99–110, 118–21

Obturator internus muscle, 142
Optical data analysis, 14–15, 175–201
Optical filtering, 174–75, 191, 194–95
Orientation of specimens, 9, 11
Osteoporosis, 173, 191

Pectoralis muscle, 29
Pelvic musculature, 34–35
Pelvis, 24, 33–36, 67–68, 111, 135, 151–55, 165, 168–74
Phalanges: of fingers, 24, 128–30, 132, 134–36, 164–68; of toes, 159, 165, 168
Phase spectrum, 178, 188
Photoelasticity, 16–17, 71, 122–44, 164–68
Photostress coating, 122–24
Piezoelectric phenomena. *See* Mechano-electric properties
Polarized light: circular, 124–25; plane, 125–26
Polarizer, 125
Prandtl's membrane analogy, 123
Principal components analysis, 37–42, 46–47, 49, 88

Principal stresses, 125, 127, 138
Principle of Saint-Venant, 127
Probability plot, 83–86

Quarter-wave plates, 125

Radiographs, of bone, 3, 11, 174–75, 190–92
Radius, 17
Regression transformation, 68
Relative clavicular length, 30–31, 64
Relative path retardation, 124–25
Resultant forces, 20–25
Reversible birefringence, 123
Rock sections, 181–84, 193–95
Rotation of axes, 39, 42, 56

Scapula, 24, 27–28, 30, 63–66, 133, 138, 142, 145–46; axillary border of, 30, 159; blade of, 140–41; "combined dimensions" of, 52–54, 56–58, 116–17, 145–51; inferior angle of, 30; "locomotor dimensions" of, 50–52, 56–58, 99–110, 118, 151, 156–57; propulsion and retraction of, 27–30, 63; "residual dimensions" of, 51–53, 56, 58, 151; rotation of, 27–30, 63, 133, 137; spine of, 138–39, 141
Segregates, 89–90, 94, 108–9
Serratus magnus muscle, 27–28, 30, 133
Sesamoid bones, 138, 141–43
Short scapular muscles, 30, 138
Shoulder, 21–24, 26–31, 34, 50–67, 75–76, 131, 148, 155–59, 168
Simplicial decomposition, 96–97

Single linkage cluster analysis, 112–15, 117–18
Skull, 37, 47, 49
Square root transformation, 82, 85–86
Strain-gauge rosettes, 122–23
Stress trajectories, 126–27, 139–41, 143
Suprascapular notch, 159
Supraspinous fossa index, 30

Talus, 160–65, 168
Telemetry, 122
Tensile forces: in bone, 135–41; in shoulder, 21–24, 29–31, 50–51, 55, 57, 137
Tests of technique, 69–76; of mathematico-biological meaning, 47; of variance differences, 73
Tibia, 192
Tomograms, 190
Transformations, 1, 13, 40–41, 43, 68, 82–86, 93, 158, 170–74, 197–99
Trapezius muscle, 26–28, 30, 133, 138
Trunk diagrams, 97, 99, 101–3, 106, 108, 119

Ulna, 17
Univariate analysis, 6, 30–31, 35–36, 81–83
Univariate normal plotting, 81–83

Variance transformation, 40–43, 70–71, 77
Vectorial representation, 12, 175
Vertebrae, 183–90, 193

Walsh transformation, 175, 197–200
Wenner-Gren Foundation for Anthropological Research, viii, 169–74
Wrist, 32, 155, 168
Wrist flexor muscles, 31–32